EXPLAINI

EXPLAINING LIFE

Allen B. Schlesinger

Creighton University

McGraw-Hill, Inc.
New York   St. Louis   San Francisco   Auckland   Bogotá
Caracas   Lisbon   London   Madrid   Mexico City   Milan   Montreal
New Delhi   San Juan   Singapore   Sydney   Tokyo   Toronto

EXPLAINING LIFE

This book is printed on acid-free paper.

2 3 4 5 6 7 8 9 0 DOC DOC 9 0 9 8 7 6 5 4

ISBN 0-07-055462-5

The editors were Kathi M. Prancan and Pam Wirt;
the production supervisor was Friederich W. Schulte.
Cover and interior collages were done by Nancy Gibson-Nash.
R. R. Donnelley & Sons Company was printer and binder.

Library of Congress Cataloging-in-Publication Data
Schlesinger, Allen B.
   Explaining life / Allen B. Schlesinger.
     p.   cm.
   Includes index.
   ISBN 0-07-055462-5 (soft)
   1. Biology. 2. Life (Biology) I. Title.
QH308.2.S35   1994
574—dc20   93-23895

Contents

PREFACE   vii

CHAPTER ONE: INTRODUCTION TO SCIENTIFIC EXPLANATION   1

Observations are influenced by cultural world views which also affect the formation of scientific hypotheses. Experimental testing of hypotheses involves comparing the predictions of explanatory models with the actual behavior of the universe. The communal and evolving aspects of scientific effort result in progressively improved explanations of Nature.

CHAPTER TWO: VIEWS OF LIFE   15

The natural history tradition examined the diversity of the living state. The ladder of life was an effort to organize and make life's complexity comprehensible. Greek natural philosophy attempted to provide a unifying explanation for life. Anatomy and physiology, from Galen to Vesalius and William Harvey, probed beneath life's surface manifestations.

CHAPTER THREE: THE GREAT CHAIN OF BEING   35

There are two models of creation: eternal stability and successive change. Mechanism contributes its explanations for origins and species formation. Linnaeus relied upon homology as the basis for his classification and taxonomy of an unchanging living world. Homology can also be used in an explanation involving transformation.

CHAPTER FOUR: HISTORICAL CHANGE: AN 18TH-CENTURY CONVICTION   51

Lamarck explained life's variations as the result of the inheritance of acquired characteristics. Opposing catastrophism, Buffon and the geologists argue for a uniformitarian succession of worlds. Interpretations of the meaning of fossils depends upon one's convictions. Hutton sees a changing world with no prospect of an end.

CHAPTER FIVE: THE CHALLENGE TO CREATION BY DESIGN   69

The teleological explanation of life insists on purposeful design and a Creator-Designer. The Darwinian explanation challenges intention in the universe. The influence of geology, economics, and political theory influenced Darwin's thinking about the natural world.

CHAPTER SIX: NATURAL SELECTION   87

The struggle for existence occurs between members of a population. Variation exists but its explanation eluded Darwin. Competition, fitness, and selection are concepts which must be understood in context. The Darwin/Wallace explanation for evolution was in terms of natural selection. Scientists question how random variations can lead to progressively complex organisms.

CHAPTER SEVEN: WHAT IS LIFE?   103

The concept of hierarchical organization assists an understanding of the chemical basis of life. The discovery of cells reveals life's functional unit. Claude Bernard's insight gives a unifying quality to an organism's various organs and activities. A hint of the future direction of research comes from 18th-century speculations about forming life from inert chemicals.

CHAPTER EIGHT: ATOMS, MOLECULES, AND THE ORIGIN OF LIFE   119

In order to comprehend chemical mechanism it is necessary to understand atoms. Electron distribution underlies the behavior of atoms and molecules. Polar covalent molecules are the key to understanding biological chemistry. Hydrogen bonds help explain the shape of very large molecules. Emergent properties are revealed through progressively more complex molecular interactions and provide a possible glimpse into the origin of life.

CHAPTER NINE: THE FIRE FROM A STAR   139

The laws of thermodynamics apply to the living world. In a universe which is running down, how can life become increasingly complex? Linking exergonic to endergonic reactions is life's response to entropy increase. Catalytic enzymes are essential to the chemistry of the living state. Enzymatic specificity plays a controlling role in the operation of metabolic pathways.

CHAPTER TEN: OF PEAS AND FRUIT FLIES   159

Life's descent from prior life is its greatest mystery. Gregor Mendel depended upon the random behavior of theoretical particles, his elements, to provide an explanation for inheritance. Mendel's First Law involves the segregation of a parent's genetic content into its reproductive cells. Dominance and recessiveness are involved in the disappearance and reappearance of traits in successive generations. The difference between genotype and phenotype is an important distinction. Mendel's Second Law underlies much of life's remarkable diversity. Sutton provides the evidence that Mendel's theoretical particles really exist. The fruit fly provides an ideal subject for genetic experimentation. T. H. Morgan's group brings Mendelian genetics to maturity.

CHAPTER ELEVEN: THE DISAPPEARANCE OF NATURE   183

Research into the nature of the gene revealed life's fundamental unity of design. Miescher's pioneering work with nuclein started 125 years ago. Levene's proposed structure for DNA seemed to eliminate it from playing a genetic role. Nucleic acids are polymers consisting of nucleotide monomers. Specificity requires complex molecular structure; proteins fit the description. Griffith discovers transformation in bacteria. Avery's lab identifies the transforming substance as DNA. The phage geneticists show the way using bacteria and viruses. Chargaff's rule and Rosalind Franklin's data are used by Watson and Crick to model DNA's structure.

CHAPTER TWELVE: THE PRIDE OF THE GARDEN   213

Genes form enzymes. DNA is a self-replicating molecule. DNA molecules encode information. The genetic code is universal. Messengers and translators are required to produce the polypeptide products. Changes in DNA underlie evolution. Embryonic development requires control of transcription and translation.

CHAPTER THIRTEEN: THE SORCERER'S APPRENTICES   241

Genetic recombination is a naturally occurring phenomenon. Bacterial research provides both basic understanding and the seeds for technological applications. Restriction enzymes and cloning techniques are the basic tools of the genetic engineer. Sequencing the human genome will be finished within this century.

CHAPTER FOURTEEN: BIOLOGICAL EXPLANATION AT THE MILLENNIUM   257

Jacob portrays science as building a representation of reality. The future path of biological research will be marked not only by Nature but also by human culture.

INDEX   271

Preface

In the first sentence of his preface to *The Logic of Life*, Nobel laureate François Jacob writes, "An age or a culture is characterized less by the extent of its knowledge than by the nature of the questions it puts forward." It has been my intention, in this book, to consider some of the many ways students of the living state have asked their questions. While the answers Nature has given are treated in some detail, it is always the questions and their motivation to which the reader's attention is directed.

The extent of the historic presentation is a departure from many books about biology. My major instructional involvement for more than 40 years has been the presentation of a college course in General Biology. The remarkable growth of fundamental understanding in the areas of molecular genetics, cellular structure and function, biochemistry, development, physiology, behavior, and ecology put pressure upon the time available to consider the historical processes which shaped the unifying patterns of thought which link the various subdisciplines of biology into a coherent whole. Over time I found myself reluctantly yielding coherence to coverage. In another sentence in his preface, Jacob warns, "Contrary to popular belief, what is important in science is as much its spirit as its product...." I have attempted to convey some of that spirit in this presentation.

It might be thought that a progressive series of discoveries and conceptions would stretch back through history connecting our present molecular explanations with the atomic theories of the ancient Greeks. The story of the development of biological ideas can be told in a selectively linear fashion, but it is unacceptable to do so. History demands that we acknowledge the coming and going of the other major views of life which have occupied the attention and earned the allegiance of generations of scientists. The investigators and their views of reality changed with the times, and their questioning reflects the cultures which produced the questioners and embodies the values of their societies. The interdependence of science and society is a major theme of this book.

Among all the explanations, Darwinian evolutionary thought holds primacy. All of the observations find coherence and relationship to one another under its unifying insight. It is for this reason that the first half of the book builds to the establishment of the selectionist interpretation for life's design. The second half of the book benefits from the attainment of this insight which permits discussion of energetics and chemical activities freed from questions of teleological intent.

Portions of Chapters Eight, Nine, and Ten are intentionally presented with much less historical treatment and research detail than is characteristic of the rest of the book. In my experience, the chemical nature of life, the role of the energy laws in metabolism, and the essential aspects of Mendelian (transmission) genetics are topics which many students find difficult to learn. This is a serious problem because our current mechanistic view of life assumes an understanding of these fundamentals. I have presented these discussions in as straightforward a manner as possible with a minimum of elaboration. My hope is that for the reader encountering these concepts for the first time, the essential issues will emerge clearly. For the student who has had some contact with the material and feels that the forest has been obscured by too many trees, it is my hope that a bare-bones presentation may provide a vantage point from which to view the more detailed accounts with an increased assurance.

I have limited the topics discussed to a selection from among the conceptual bases upon which present biological explanation rests. In any process of choice there is bias. Mine is influenced by an affinity for the cellular and molecular levels of biological organization. Considerably more of this book's attention is devoted to genes and gene processes than to the diversity of organisms, for example. This is not entirely an arbitrary decision, however. Genes came before structure, and they underlie what we perceive as anatomy, life cycles, and behavior. Genes, along with certain chemical and physical realities, lie below the surface of life's manifestations. The search for the universal qualities shared by all living organisms is the central intention of the book.

The influences upon a teacher of science reach to the earliest schooling conversations and readings and in the course of an extended career are blended so that individual contributions may become difficult to identify. Some influences are so distinct and influential that they remain isolable. Among teachers and investigators, Paul F. Brandwein, H. Burr Steinbach, and Fred Smith provided example and inspiration. My attitude toward the process of science, particularly its being embedded in the human experience, reflects the writings of François Jacob and Sir Peter Medewar and expresses the historical perceptions of William Coleman and Garland Allen. The full complexity of the relationship between science and culture is revealed most brilliantly by Horace Freeland Judson's synthesis of persons and their intellectual products in the history of molecular biology.

More than all of these influences combined is the effect of the dynamic that is established between a teacher and his students. I have been most fortunate in having been involved in a synergism with students at Creighton University in which I have been stretched to lengths I had not anticipated in my efforts to fulfill their expectations.

The production of this book was facilitated by the constant and patient guidance in computer use and publication design provided by Bob Guthrie.

My wife, Julie Schlesinger, provided both enthusiastic support and incisive criticism. I dedicate this work to her with love.

Allen B. Schlesinger

EXPLAINING LIFE

CHAPTER ONE

INTRODUCTION TO SCIENTIFIC EXPLANATION

"All science is the search for unity in hidden likenesses."
— J. Bronowski

Explanations

Science, as well as much of human thought, is about explanation. There may have been cultures which accepted the world without any effort to make it comprehensible, but history informs us that human nature requires more than acceptance of reality. Science is one of the processes of thought by which we have attempted to mentally grasp the world. This is an interesting line of thought because it makes a very important assumption. If we embark on the task of explaining the world we must have assumed it is a knowable place. Science affirms a belief that the world is indeed knowable because it is orderly, that it follows a set of unchanging laws, and that these laws can be understood by the human intellect. This assumption underlies the attitude of Western civilization toward Nature.

There are individuals who believe the world is chaotic, unpredictable, and therefore unknowable. There are others who believe that the world is knowable but is essentially mysterious and magical and therefore unknowable unless power over the magic can be gained. Most of us have a mixture of convictions about this subject, believing that some aspects of the world are quite knowable (the rotating earth as an explanation for the highly predictable alternations of day and night) but that others are absolutely unexplainable (the early death of wonderfully talented and good people, for example).

There have always been explainers and their explanations. Every culture has explanations for the origin of the world, the patterns of the seasons, the reasons for death, the presence of evil, and the nature of truth. How can the quality of an explanation be judged? How do we decide if an explanation is valid or invalid?

We do not judge art or music with the same standards that we apply to scientific work. Each way of knowing and each kind of explanation is judged by its own set of criteria. There is, however, one universal test that all explanations must pass. Michelangelo's explanations in marble, the emotional depth of a Beethoven symphony, Homer's stories of the mating of Gods, and Einstein's chalkboard formulas all share the common quality of having *satisfied* the intended audience. Don't think that satisfaction with an explanation is a superficial judgment. To be *satisfied* with an explanation as to the way the world came into existence or why we cannot live forever goes well beyond casual acceptance. When we are *satisfied* with such an explanation, we have put it to the test of fitting into a construct of our experience, our values, and our aspirations.

Scientific Observations Aren't Totally Objective

In this book we will examine *scientific* explanations. Specifically, we will be examining the explanations *scientists* have constructed concerning the living state. We will be applying a different set of judgments concerning satisfaction than we would be if this were a book examining those explanations of life created by poets or philosophers. Werner Heisenberg, a physicist, wrote, "What we observe is not Nature itself, but Nature exposed to our method of questioning." This is a profound insight, and it is worth a little further consideration.

Most of us first learned about science in grade school, typically beginning as early as first grade when we made a leaf collection or were taken to the museum and were awed by the dinosaur skeleton. We associate science with rocks, stars, bones, and blood. In other words, science is about *things*. We were told that scientists are particularly precise people, making precise observations and precise measurements. Another characteristic of science, we were told, is its insistence upon objectivity: personal opinions are not allowed to color the scientist's observations and measurements. This early exposure to science flavors our adult attitudes. We retain a vision of a precise, objective, thing-oriented activity accomplished by scrupulously unbiased technicians. Is there such a thing as truly "objective" observation?

"All science involves a selection process. No scientist observes everything."

All science involves a selection process. No scientist observes everything. There is always a decision as to what to observe. This is not a superficial statement. In the process of deciding what to observe there is obviously a decision not to observe other things. A first-grader's leaf collection is highly edited; only intact, colorful, "interesting" leaves will be pasted on the chart. Why doesn't the child collect broken leaves? Aren't broken leaves actually present in abundance? Why include only the colorful ones? Aren't the drab leaves actually the most numerous? The eye sees all kinds of leaves but the brain makes decisions as to which ones are to be collected. Nature "itself" is not really represented very accurately in the collection. The child teaches us what Heisenberg is talking about when he says that "what we observe is not Nature itself" but our *selected vision* of Nature. We decide which aspects of a situation to examine, and we decide how to go about the process. We frame the questions to be asked. Nature can only respond to the particular question asked.

Science Is Identified by the Kinds of Questions It Asks

Both myths and science attempt to provide us with representations of the world, and they describe the forces which control nature and the affairs of humanity. It is very difficult, however, to test a mythic explanation which involves the intentions of a goddess or the powers of a pure heart. It is entirely possible that mythic powers exist. The fact that we cannot ask for demonstrations of them in the laboratory does nothing to deny their reality. For the scientist, however, what is not accessible for questioning is not very informative. What is it that informs the scientist? What kinds of questions does science ask?

François Jacob helps us understand the scientific intention when he states, "For science, there are many possible worlds; but the interesting one is the world that exists and has already shown itself to be at work for a long time. Science attempts to confront the possible with the actual. It is the means devised to build a representation of the world that comes ever closer to what we call reality."

"Science attempts to confront the possible with the actual. It is the means devised to build a representation of the world that comes ever closer to what we call reality."

In confronting the possible with the actual we ask a question which permits the world to respond as much as possible with its *own* actions. If we ask, "What is life?" Nature cannot respond with *its* activities. A philosopher can ask that question and then create a response *for* Nature from an inner vision. A scientific question might be, "Is life possible without oxygen?" That's a question to which Nature can respond directly. All the investigator needs to do is place a living organism under a set of conditions where oxygen has been excluded. If all the other environmental conditions are met (energy source, water, proper temperature, etc.) Nature will deliver a response — either yes, the organism lives, or no, the organism dies. Nature will tell us that *some* organisms, certain bacteria, for example, can live without oxygen. For the majority of species the answer will be no.

The first thing we notice about the scientific kind of question is that it elicits answers which are much less comprehensive than the philosophical. The kinds of questions scientists ask of Nature reveal *pieces* of reality. It takes a lot of skill to assemble many scientific questions, each one eliciting only a piece of information, so that ultimately a "representation of the world" can be constructed. A good deal of this book will be devoted to examining the kinds of questions that scientists have asked about the nature of the living state. Some of these, I think you will agree, are extremely ingenious.

Science Is Affected by the Culture Around It

The belief that scientists go out and observe Nature with total objectivity and that somehow Nature reveals itself to them couldn't be further from the truth. What one knows and what one feels to be significant will *influence* the type of questions one asks of Nature. Here's an example of this intellectual bias. During the 18th century the industrial revolution changed the nature of Western culture in many ways. Not only were there changes in the social and economic structure of Europe but also in the kinds of questions which scientists asked of Nature. The steam engine was the device which powered the industrial revolution, and coal was its primary energy source. As the technology of coal burning (combustion) became more sophisticated it became possible to accurately measure the energy content of various kinds of fuel. Calorimetry, the measurement of the heat content of substances, was widely used. It was possible to measure the heat content of coal, wood, or literally anything at all.

It was in this intellectual environment that Antoine Lavoisier first proposed a combustion theory of animal physiology. He asked, "Is Life a Process of Combustion?" Lavoisier burned candles in his "calorimeter" and measured the amount of heat given off by a given weight of wax. He then placed small animals in the same device, allowed them to respire, and measured their heat production. By weighing the animals before and after a period of heat measurement, Lavoisier was able to calculate the heat given off by a given amount of living tissue. The heat production of a given mass of burning candle and that of a respiring mouse were remarkably similar. Lavoisier concluded that life was essentially a combustion process. Scientific thinking is a product of its time and place in exactly the same way that art and philosophy reflect the societies in which they occur.

"Scientific thinking is a product of its time and place in exactly the same way that art and philosophy reflect the societies in which they occur."

The point of this discussion has been to emphasize that science is a human endeavor flavored by the attitudes, interests, and values of the persons involved. It is not a disembodied accumulation of facts.

The Formation of Scientific Concepts

Albert Einstein provides us with a deep insight into what

is actually going on as a scientist goes about the task of asking questions of Nature. "It seems," he wrote, "that the human mind has first to construct forms, independently, before we can find them in things." We've all had the experience of being shown inkblots and seeing in them birds or faces or some other object. Sometimes we've been embarrassed to state what we see. Einstein points out that we must have the "form" in our minds *first* before we can find it in the inkblots. Imagine if you will an ink blot which is a precise outline of a fish. Is it possible that a person who has never seen a fish and therefore has no mental image of one could, upon being shown this blot, announce that it is a fish? Such a person would say that the blot is either nothing at all, or perhaps that it is a badly drawn baseball bat. In other words, something must first be a *mental* concept (Einstein used the word "form") before one can "see" it out in the world. Lavoisier had the conception of a combustion engine in his mind. He had seen coal-burning engines performing work. This mental construct (or model) allowed him to *imagine* a living organism as a combustion engine. So Lavoisier transferred an *image* from his mind to a *thing* in the world.

Transfer is not the only thing Lavoisier did. He also *transformed* the image. A living thing isn't actually a steam engine. We have a word for this kind of transformation. We say that the steam engine is a **metaphor** for the energy conversions in the living organism. The use of metaphors is very useful in explanations since we can go from something we know quite well (the steam engine) to something we are uncertain about (the organism) with a sense of familiarity. The danger in their use is that we may forget that metaphors are only temporary scaffolds for our thoughts. We may come to depend so much upon the scaffold that we confuse it with the building we are trying to erect. The critical thing is the selection of a metaphor that will help us with our construction because it is truly applicable.

Comparisons Underlie Scientific Experimentation

How could Lavoisier have known if the combustion metaphor really applies? Was he correct in making this transfer? Einstein helped us see how the next step occurs.

"There is a process in which the 'facts of observation' must be compared with the mental conception."

"Knowledge cannot spring from experience alone but only from a *comparison* of the inventions of the intellect with the facts of observation." Lavoisier would have to have compared the combustion engine model with the way in which the animal actually produces heat. There is a process in which the "facts of observation" must be compared with the mental conception. This, for most nonscientists, is the most confusing aspect of science. *This process of comparison is called an "experiment."*

In everyday speech we call a variety of activities "experiments." When we were children we mixed several things together, perhaps some vinegar, some sugar, some salt and performed "an experiment." Or we were told by a teacher that he or she is "experimenting" by using a new way of explaining a complex lesson. These are not experiments. An experiment, as Einstein indicated, is a comparison. Something out in the world will be manipulated, and the outcome of the manipulation will be compared with the outcome as predicted by the mental model. In Lavoisier's case, the mental model was a burning process. Fuel, when burned, produced heat. The amount of heat should be proportional to the amount of fuel. A candle weighing one ounce should produce twice as much heat as a candle weighing half an ounce, assuming the candles are made of the same kind of wax. Lavoisier actually burned candles and measured the heat production. So far so good. Now for the animal. If an animal is actually burning fuel, it should produce heat, and furthermore the heat produced should be proportional to the amount of fuel (animal weight) which disappears. If the process is *really* combustion, the amount of heat produced and the amount of fuel lost by an animal should be very similar to that Lavoisier found for the candles. He was gratified when he examined the figures. His mental model of combustion predicted certain outcomes, and the "facts of observation" (the data from his experiments) when *compared* with the model's predictions supported his imaginative idea. He questioned Nature with great insight and the answer he received opened up a view of life's workings never before available.

"If the process is really combustion, the amount of heat produced and the amount of fuel lost by an animal should be very similar to that he found in the candles."

This is a good time to recall François Jacob's statement that "Science attempts to confront the possible with the actual." This is exactly what an experiment is: a confrontation. The scientist forms a "possible" mental model. In formal terminology, the scientist has stated a **hypothesis**. It is an idea about the world and the way it operates. Next comes the confrontation — the comparison of the hypothesis with the way the world *actually* works. This is an **experiment**. If we keep Jacob's statement in mind when we examine the details of experiments we will find it a great deal easier to follow the intention. The scientist is in a confrontation, that of the possible explanation of the mind against the actual behavior of the world.

Science Is a Communal Activity

Then isn't science basically a fragmented collection of thoughts about reality supported by individual experimental verifications? It is more than that because of some very fortunate qualities which emerge from the nature of science and the way it is practiced.

In science there is the obligation to speak in specifics. If a candle is burned it must be a real candle, of a certain wax content, actually burned, and weighed under clearly stated conditions. There can be no veiled meanings, no unrevealed circumstances. This is the basic attribute of science. A scientist's creativity must be evident and accessible to others.

In practice, scientific findings are revealed through publication in **scientific journals**. One individual's "possible" (the hypothesis) is stated, and the results of the (experimental) confrontation with the world's "actual" can be read and judged by the entire scientific community. If the work is judged as being significant many other scientists will repeat the experiment. Their results will be compared with the original, and there will be another confrontation — this time between scientists who may get differing results. In time, the hypothesis will be judged either valid or invalid. This aspect of science results in a communal vision rather than an individual one. It is from this communal aspect that science draws its power to be convincingly correct. Science is a coherent construction rather than a collection.

The greatest strength of scientific thought lies in the ability of all scientists to participate in what is truly a single, shared intellectual construction. There exists a web of thought with connections and interdependencies that provides two qualities which are unique among all forms of human activity. The first is the easier to communicate. Science successfully crosses cultural lines. Although various cultures may have differing moral values and philosophical convictions, if scientists examine a hypothesis using the *same* experimental procedures Nature will deliver the same outcome regardless of the desires or convictions of the persons who are involved. No matter what you believe about water's sanctity, healing powers, or origin at creation, water will freeze at 0°C, boil at 100°C, and dissolve salt but not fat.

The *universality* of science is easy to comprehend. A more difficult aspect of science is its *internal consistency*. Imagine 100 poets working independently of one another, each creating a personal statement. To make this circumstance truly accurate, imagine that each of these poets lives at a different time, in a different country. Your challenge is to fit their poems into one coherent statement. You may not change their words in any substantive way. It is very unlikely that the collected poems would have any truly meaningful relationship to one another. It would be a collection but it would not tell a coherent story. In science that is *not* the circumstance; individual scientists all over the world living in different centuries have questioned Nature by using the same technique of hypothesis formation and experimental testing. The pieces are coherent. The facts concerning water's freezing and boiling points, and its ability to dissolve some substances and not others, fit with the roles water plays in biological systems. When chemists determine that water consists of two hydrogen atoms and one oxygen atom the facts about these atoms fit precisely with the properties of water.

"None of the individual explanations can be considered valid if they will not relate to the totality *of the scientific explanation."*

This is *internal consistency*. None of the individual explanations can be considered valid if they do not relate to the *totality* of the scientific explanation. Science is doubly demanding. Its first demand is that the individual hypothesis be tested with great experimental rigor. After surviving this test, the new explanatory piece must be able to fit smoothly into the existent scientific understanding. Sometimes, however, what seems to be an accurate piece simply will not fit.

Science Evolves

Imagine yourself doing a crossword puzzle. You've worked out a few words to your satisfaction until you attempt to fit the next word, which you are absolutely certain is correct, and it obviously will not fit. You hesitate to dismantle the words you've already put in position, and your tendency is to reject the new word. If two or three words have already fit with one another (have internal consistency) you are reluctant to give up the progress you've made. If you are honest about the puzzle you have to at least consider the possibility that you may have gotten a fit with the wrong individual words.

Science is prone to this kind of error. There have been constructions of explanations which turned out not to be able to accept a clearly correct new piece of information. This is deeply disturbing both to the individual scientists who worked on the construction and to the scientific community which had confidence in the process. But if the new piece of evidence clearly should fit and simply won't then the construction must be modified. What typically can be seen in hindsight is that pieces of information had been assembled before a critical piece was available. The pieces fit well enough until the whole assemblage was asked to accept the critical newcomer. This calls for the disassembling and reconstruction of the concept. Recall the words of François Jacob. "It [science] is the means devised to build a representation of the world that comes ever closer to what we call reality."

Science comes ever closer, but it never reaches an end. New pieces are devised, and the fit becomes increasingly more precise. Now and then a poorly fitted portion must be redesigned, its pieces reassembled so as to better accept new information and allow the entire construction to preserve its internal consistency. Because science has been in existence for a long time and has received so much human effort, the edifice is both complex and strong. A new finding which fits validly has a very high probability of being correct.

Scientists are content to work on the construction. They do not demand completion. The anticipation that comes from devising and testing a new explanatory model and

"Science long ago abandoned the belief that it would attain absolute certainty."

the gratification that follows if it slips smoothly into position in the massive construction that exists and has been growing over the centuries is very much like the building of the great medieval cathedrals. Individual artisans worked at specialized trades cutting stone, forming timbers, casting bells, etc. Errors were made, steeples collapsed, and individual lives were lived in doing not accomplishing. The comparison is not accurate since the cathedrals were eventually completed. As we have seen, science aspires to "come ever closer to what we call reality." Science long ago abandoned the belief that it would attain absolute certainty.

Does Science Assure Truth?

Science shares some basic qualities with most other systems of thought. It seeks to detect patterns of order in the subjects it explores. Scientific explanations, like other explanations of the world, are human mental creations. We have stressed that it is in the method of testing the "possible" against the "actual" that science has some unique qualities.

Does science guarantee "truth"? In a very limited sense the answer is yes. For example, chemists can speak with total confidence that if one mixes certain proportions of certain elements under specified conditions a given reaction will occur. This capacity to *forecast* is very impressive. The ability of science to predict the location of a satellite landing on Venus years after launch on earth should impress even the least technically talented person. The ability to *predict* with precision implies that the person capable of making the prediction must truly *know* the nature of the processes involved. There are no other forms of human thought that have the predictive capability that science has demonstrated so convincingly. Isn't this precision a clear indication that science permits the acquisition of truth? This is, indeed a *part* of possessing the truth, but let's look at another aspect of prediction.

Some 500 years before the birth of Jesus Christ the Babylonians were perfecting their ability to predict various astronomical events. Their skills at prediction were based upon extraordinarily complete observations and meticulous record keeping. Their tables and charts gave them the ability to recognize patterns which repeated at definite intervals, and it was this kind of prediction at which the Babylonians were successful. The Babylonians did not, however, have any significant concepts which *explained* the motions of the planets. They had no ideas about the physical nature of the universe. Oddly enough, they do not seem to have been at all interested in speculating about the physical nature of the heavens with which they were so familiar. Is it important to understand the underlying reasons for repetitious events? Isn't the ability to predict with precision enough? In other words, isn't a *piece* of truth completely true as far as it goes?

The Babylonians were fascinated with the repeating patterns and cycles in the heavens, and they tried to extend their predictive capabilities to other events. They attempted to predict earthquakes and the appearance of plagues of insects. They were clearly less successful with these kinds of events. Why do I say, "clearly"? Because we have since learned that the motions of planets and the appearances of locust hordes are entirely different kinds of processes operating under entirely different causes. The Babylonians were unable to extend the truth of their celestial charts because they had no basis upon which to judge the nature of the movements they observed. One indicator of explanatory truth is its capacity for *extension*. If we've discovered a really *fundamental* quality about a chemical, for example, we should find that quality in other chemicals in some recognizable form.

"One indicator of explanatory truth is its capacity for extension. *If we've discovered a really* fundamental *quality about a chemical, for example, we should find that quality in other chemicals in some recognizable form."*

The reason that this subject is important is that science is accepted as being a very successful predictor. It has also had very impressive successes as an explainer. Think about the germ theory of disease as an example. Other forms of thought had provided explanations which ranged from punishment for sinful behavior to spells cast by witches to the effects of "bad" air. With the identification of bacteria and their transmission from one individual to another, we were in a position to take effective action to prevent the spread of what had been totally mysterious afflictions. Nuclear reactions, genetic engineering, antibiotics, weed-killers, pesticides, and all the other scientific contributions to technology are the result of the explanatory capabilities of science.

The success of the scientific thought process in topics where it is highly effective (physics, astronomy, chemistry, biology) tempts us to perceive science as a truth-generating machine. Let us never forget that the process starts with a person asking a question. We've become very skilled in asking physical questions about the nature of matter and the ways in which energy flows. As a result we have what scientists like to call "robust" (sturdy, vigorous) explanations which permit truly remarkable predicitive accomplishments. I am typing these words on a computer keyboard. The computer chips, the software program, and the storage disks are all expressions of the combined power of explanative understanding and predictive utility. Underlying all of this is thousands of years of human questioning. There have been countless questions which were phrased badly, and Nature was unable to respond with meaning. The physical sciences have been successful because the questions we have learned how to ask have permitted Nature to respond with increasing revelations.

Can the scientific way of knowing be extended universally to all areas of reality? A thoughtful response centers upon two things. First, since it is we who must phrase the question, do we know what questions to ask in all areas of reality? Are we in possession of sufficient information to question Nature profitably? I remind you of the fact that until you know that bacteria exist you cannot possibly ask if they cause disease. Second, and much more difficult to think about, is this. If I can imagine something, such as greed, does it *exist* in Nature in a manner which permits scientific questioning? We will all insist that we can recognize greed when we see it, but do you think you can phrase a scientifically meaningful question about it?

Science is a tension-filled enterprise. On one hand it is imaginative and has the exuberance of confident creativity. Like all creative people, scientists cannot abide regulations and restrictions. They join with artists in the vision of limitless expression of human wonder. Yet, unlike artists, scientists function under a self-imposed restriction. Scientists *must* submit their imaginative possible to the judgment of the actual. If *Nature's* response denies their creation scientists *must* abandon what all of us prize so highly — being right.

CHAPTER TWO

VIEWS OF LIFE

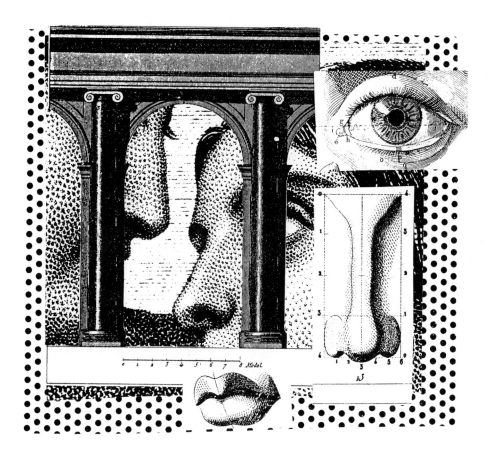

"Life can only be understood backward but it must be lived forward."
— Soren Kierkegaard

This chapter will examine some of the changing views of life over the more than 2500-year span between the early Greek efforts and the close of the 18th century. We will be emphasizing a selection of thought patterns, not dates or names. Thoughts are not as tidy as calendars; they do not cease to have meaning when pages are torn off. Some ideas flow into later centuries, others disappear for hundreds of years only to reappear in slightly different form to resume vigorous participation. Some will sound silly, and you may wonder why we are considering them at all. I tell you this at the outset in an attempt to avoid the concern most of us experience when a "story" doesn't progress, in precise chronological order, to a tidy conclusion. The kinds of questions asked of life over this great span of time reveal a great deal about the values and hopes of the cultures involved. We will see that until the kinds of questions changed dramatically, life would remain hidden beneath its surface appearances.

The Study of Things Compared with the Study of Laws

For most of the period from Greek antiquity to the beginning of the 19th century, questions asked about life centered on the "integral" organism. An individual, entire, functioning animal or plant is the "natural" object of our attention. Even when we consider organisms in groups, we maintain a persistent vision of individuals in an assembly. By centering attention on individual organisms — animals and plants as units to be observed, named, described, and placed into categories — the vision of the living world took on a different character as compared with the inert world being explored by the physical sciences. Individual raindrops do not capture our attention in the way individual sparrows do.

"For most of its history biology emphasized the qualities of individual creatures. It was only in the 19th century that life, as a state of existence not dependent upon the qualities of any individual organism, began to be studied from the point of view and with the techniques that had guided the development of the physical sciences."

For physics and astronomy the world was seen as one of design operating under the direction of as yet unknown laws. These laws were knowable, however, and it was their discovery which the physical scientist pursued. In studying acceleration, for example, it doesn't matter a great deal whether the ball which is rolled down an inclined plane is made of wood or ivory. *Acceleration* is a conceptual abstraction. The balls are devices to aid in the process of understanding the abstraction. Life, however, was not studied as an abstraction. Individual living things were not seen as *representations* of life but as life itself. In the analogy given, if it is believed that it is the kind of ball which determines the rate of rolling it would be natural to place a great deal of attention on the specific qualities of the wood and ivory. For most of its history biology emphasized the qualities of individual creatures. It was only in the 19th century that life, as a state of existence not dependent upon the qualities of any individual organism, began to be studied from the point of view and with the techniques that had guided the development of the physical sciences.

Natural History

Prior to the 19th century there were three traditions which

provided the basis for our understanding of the living state. Although it is convenient to treat each of these separately, it is absolutely essential to keep in mind that they were actually blended into one another and became intellectually inseparable.

The oldest was what came to be called **natural history**. This started off as a collection of observations concerning all of the world, living and nonliving. As an outgrowth of a hunter-gatherer society's interests, its earliest phases stressed the descriptions of organisms, noted where they were found, and documented the lives that they lived. The *external appearances* of organisms dominated early natural history since its interests were well served by the readily visible. With increasing knowledge about an ever-enlarging list of creatures, **nomenclature** (naming) and **taxonomy** (classification) were natural developments. Identifying an organism by name isn't academic in the case of poisonous mushrooms. Putting organisms into meaningful groupings is a natural human inclination and, as we shall discover, the arguments as to what is truly a meaningful grouping can rage for centuries.

Aristotelian Natural History

Aristotle (384-322 B.C.) brought natural history to a remarkable level of maturity through extensive observations of organisms in their natural environments. His interests are indicated by four major surviving works titled: *Natural History of Animals, On the Parts of Animals, On the Generation of Animals,* and *On the Psyche.*

Diversity

What is it that Aristotle valued? Clearly he had a remarkable capacity for observation. His descriptions of the organisms, their structures and activities, and their distribution from place to place point to a conviction that explanations of life were to be based upon the variety of living forms. This focus on **diversity,** noting the remarkable ways in which life swam and crawled and ran — detailing the appearances and structures of its endless variations and recounting the marvelous ways in which it survived and flourished — comes through to us and

*"Just what did these
natural historians
have in mind? They
wanted nothing less
than the accom-
plishment of an
impossible hope:
the complete knowl-
edge of every living
thing on the planet.
They could only
comprehend life by
knowing each of the
creatures."*

*"Western culture
has molded our
thinking to antici-
pate that everything
we encounter has
some kind of pattern
and we attempt to
find some indication
of order in even the
most hapahazard
appearances. When
we ask questions of
Nature we are
evidencing our
conviction that
there is a hidden
pattern of meaning
beneath the sur-
face."*

leaves no doubt as to Aristotle's fundamental conviction. It is only by knowing each one of life's creatures that an overall understanding can be accomplished. We see this conviction throughout most of the natural history tradition of the study of life. The great majority of students of the living state were observers and collectors of life in its enormous diversity. Our museums preserve only a tiny fraction of the collections which resulted from centuries of exploration, collection, description, and classification.

Just what did these natural historians have in mind? They wanted nothing less than the accomplishment of an impossible hope: the complete knowledge of every living thing on the planet. They could comprehend life only by knowing each of the creatures.

Unity

There is a practical problem which arises as the list of known creatures becomes increasingly large. We seem, as a race, to have an urge to look for evidences of **unity** amid even the most diverse assemblages. This search for unifying identities is more than a simple effort at convenience. Western culture has molded our thinking to anticipate that everything we encounter has some kind of pattern and we attempt to find some indication of order in even the most hapahazard appearances. When we ask questions of Nature we are evidencing our conviction that there is a hidden pattern of meaning beneath the surface.

The earliest efforts to identify hidden meaning in the different kinds of creatures were efforts at classification. Each age and culture sought to group organisms according to those characteristics which were meaningful to that culture. Other than practical differences such as poisonous and nonpoisonous snakes or toxic and nontoxic mushrooms it was by no means obvious which characteristics were to be selected as the basis for classification. For example, if all four-legged creatures were grouped, horses and turtles would qualify as quadrupeds and two-legged humans and birds would be linked as bipeds. Aristotle's decision to separate oviparous (egg-laying) from viviparous (giving birth to live young) animals permitted him to separate horses from turtles and humans from birds and, in a remarkable piece of insight, fish from whales.

All of his judgments were based on observable appearances, and it was this approach which would dominate natural history for the next 2000 years.

Complexity and the Ladder of Life

It was Plato, Aristotle's teacher, who formalized the idea that some creatures were "lower" and some were "higher." In discussing "life" with students, I have never met a person who genuinely believed that there is no fundamental difference between a slug and an eagle. That's because the students I know accept a basic Platonic concept, namely that complexity of organization indicates a hierarchical system with rankings which are natural and unarguable.

To give a concrete example, a single individual person has all of the basic attributes of our race. If we have a group of such persons, a family for instance, each of the persons retains the basic human qualities, but the assemblage takes on some additional ones. Love, loyalty, rivalry, and respect stem from the fact that there is an *interaction* among persons that cannot occur when only one person exists. If we move to a community composed of many families there is a set of qualities which would never have been revealed had we not moved to levels of increasing complexity. The expression of the full potential of human nature requires increasingly complex environments.

So it is with organisms in their natural world. It was the natural history heritage which verified the view that all of life was not equally complex in its structure and its behavior. Plato's *scala naturae* (ladder of nature) formalized the obvious: some creatures are more complex in their organization than others. It was the efforts of thousands of naturalists which created the systems of classification which grouped organisms into meaningful arrangements. Typically the simpler creatures, those with less obvious complexity of design, were portrayed at the bottom of the organizational chart and the more complex ones, like ourselves, were placed in ever ascending order. Of course this is chauvinistic. If Plato had been an

"If Plato had been an amoeba the charts would have been reversed. Simplicity would have been the ultimate attribute."

amoeba the charts would have been reversed. Simplicity would have been the ultimate attribute.

The Species Problem

The arrangement of organisms into patterns presents a challenge to the one doing the arranging. Natural history grappled with what has come to be called "the species problem." If organisms were to grouped into "kinds" what were the rules for deciding when a newly encountered organism properly belonged in a known group and when would it be appropriate to start a new "kind"? Lurking behind what some of us would call petty arguments over "academic" distinctions was an issue which touches one of the the most sensitive nerves in the human psyche. Regardless of where the *edges* may be, *kinds* of creatures clearly exist. The borders between species of sparrows may be arguable but not the existence of ducks and hummingbirds. How did the kinds of creatures come into existence? Or, more correctly phrased, how did *we* come into existence? Consideration of the clash between the alternative explanations for life's diverse manifestations, its many kinds of organisms, is best delayed until we have prepared the stage for the confrontation. The natural history tradition was a major contributor to evolutionary thought.

Medicine

A second source of information and speculation about life is **medicine**. Since living things were observed to sometimes recover from disease and injury it was inevitable that humans would attempt to increase the probabilities for healing. Primitive efforts resulted in a heritage of herbs, charms, prayers, and practices. Much of early medicine is a reflection of the values and convictions of the cultures involved. Appeal to the forces in nature, confidence in the spiritual possibilities for healing, and dependence upon religious mysteries are not foolish to a culture in which absolute power has been assigned to the gods. We sometimes ask if the ancients really "believed" that wearing an amulet would prevent disease. I imagine that their attitude was very much like ours with regard to being vaccinated. They, too, wanted to have protection from disease provided by a power which was respected by the best minds in the community.

Anatomy

In 1543, Nicolaus Copernicus described the motion of the planets as they circled the sun in *De Revolutionibus Orbium Coelestium* and Andreas Vesalius provided the first comprehensively accurate description of the human body in *De Humani Corporis Fabrica*. The external universe and the universe within were both coming into sharper focus in the middle years of the 16th century. The history of the prior 2000 years was quite dissimilar for these two views of reality.

"We, as a race, had charted the stars while our interior universe remained unexplored and unmapped."

While the heavens were being studied with remarkable precision by primitive societies all over the world thousands of years before the ancient Greeks, even the ancient Egyptians with their fascination with body preparation for mummification used the same word to designate nerves, muscles, arteries, and veins. Primitive societies were able to forecast eclipses, navigate over open oceans by using celestial guidance, and devise complex calendars by using a variety of planetary and stellar information. We, as a race, had charted the stars while our interior universe remained unexplored and unmapped.

It was the Greeks who, about 300 B.C., undertook systematic dissection of the human body. For approximately 500 years there was a tradition of anatomical investigation so that by the 2nd century A.D. Galen could codify anatomical understanding which was to remain essentially unchanged for 1500 years.

For reasons which cannot be fully verified, systematic study by dissection declined following the 2nd century. It is clear that dissection of the human body was no longer considered to be an appropriate academic pursuit but whether this was due to religious proscription or a shift in interest away from the fabric of life remains debatable.

What is confusing, to most modern readers, about both ancient and much of later European medicine, is its blending of fact (typically anatomical and symptomatic) with what we see as unfounded speculation as to function and cause. We are willing to accept speculative statements but we want them to be identified as such. It is the blended acceptance of what we consider to be valid with unsubstantiated thought that the modern mind finds disturbing. It is informative to consider an example before attempting to clarify our rejection of an earlier way of knowing.

A Look into the Inner World

Galen believed that blood on the right side of the heart was mixed with air from the lungs on the left side to form the absolutely essential vital spirits or *pneuma*. There is a partition (the *septum*) between the right and left ventricles, the two larger chambers of the heart. Since most of us have little reason to dispute Galen (since we cannot accurately recall the circulatory system as we learned it in high school) the only strange-sounding part of all this is the mystical sounding "formation of the pneuma." But how was the blood supposed to get through the *septum* so as to mix with the air from the lungs? Galen solved this problem by speculating that there were invisibly small pores in the *septum* and that the blood passed through them to accomplish its purpose. To our modern way of thinking, these speculative pores are an insurmountable problem. We are very judgmental about Galen invoking invisible pores, but we are quite content with today's scientists using explanations involving invisible electrons. As we shall see, this *selective* acceptance of evidence lies at the heart of our finding fault with the speculative aspects of early medicine and all of early science. This is a matter of having confidence in one's sources of information. Values are a very important part of scientific confidence building.

"We are very judgmental about Galen invoking invisible pores, but we are quite content with today's scientists using explanations involving invisible electrons."

What happened to all the blood which came from the liver and which was assumed to mix with the air? Galen's theory assumed that it was consumed in the formation of *pneuma* and had to be constantly replaced by new blood produced by the liver. In Galen's thinking it was the passage of *pneuma* to the various organs of the body which was the essential feature; blood was an expendable commodity very much like the air which we breathe in and out so abundantly.

For over 1500 years the Galenic school of thought dominated the practice of medicine. There was sporadic dissection of the human body at the medieval universities during the 14th and 15th centuries, but little new information was obtained. It was not until the 16th century that Andreas Vesalius (1515-1564) undertook the reorganization of anatomical thinking.

De Humani Corporis Fabrica

The clandestine obtaining of bodies for dissection has been portrayed in literature and folklore, but the real feel for this activity has never been more strikingly communicated than in the words of Vesalius himself. In describing the way he recovered the body of a man who had been partially burned and then tied to a stake he wrote:

> I climbed the stake and pulled the femur away from the hipbone. Upon my tugging, the scapula with the arms and hands also came away. After I has surreptitiously brought the legs and arms home on successive trips... I allowed myself to be shut out of the city in the evening so that I might obtain the thorax which was held securely by a chain. So great was my desire to possess these bones that in the middle of the night, alone and in the midst of the corpses, I climbed the stake with considerable effort and did not hesitate to snatch away that which I so desired.... I carried them some distance away and concealed them until the following day when I was able to fetch them home bit by bit through another gate in the city.

In 1543 Vesalius published his book, *De Humani Corporis Fabrica* (On the Structure of the Human Body). This monumental effort brought together the accumulated facts and speculations of thousands of years and presented them in the light of Vesalius's detailed examination. It is instructive to learn that much of Galen's errors went uncorrected by Vesalius who was caught between his confidence in his own observations and his respect for authority. In no place is this more clearly demonstrated than in his statement concerning the Galenic theory of the pores in the *septum*. In speaking of the pitted appearance of the *septum* he tells us that in an examination of the pits, none:

> so far at least as can be perceived by the senses,

penetrates through from the right to the left ventricle, so that we are driven to wonder at the handiwork of God, by means of which the blood sweats from the right into the left ventricle through passages which escape human vision.

It has been suggested by historians of science that Vesalius was being sarcastic. To support this view, they point out that in the second edition of the book Vesalius returned to this issue and wrote, "However much the pits may be apparent, yet none, as far as can be comprehended by the senses, passes through the heart from the right ventricle into the left." This seems to be as direct as Vesalius can bring himself to be in confronting traditional teaching.

This is the point I raised some time back when I pointed out that the modern mind has difficulty accepting the way medieval and renaissance scientists were able to blend their acceptance of traditional authority with the evidence supplied by their own more informed investigations. What puzzles us is their inability to separately evaluate the two sources of knowledge. We do not have a problem because we have succeeded in separating our various values from one another. We draw great comfort and inspiration from our myths and sacred writings, but most of us do not consider these sources to be applicable to scientific modes of thought. The history of science is, to a very great degree, an account of how this separation came about.

Although Vesalius did not abandon a degree of dependence upon traditional authority, he respectfully corrected a large number of major anatomical errors. He was much less restrained in addressing popular beliefs and superstitions. There are equal numbers of ribs in men and women, he pointed out, regardless of biblical indications that men should possess one fewer. He dismissed legends concerning such things as bones in the heart and a mystically indestructable bone at the base of the spine from which the body would be regenerated on Judgment Day.

His illustrations were striking. Prior to *De Humani Corpus Fabrica,* drawings of anatomy were almost exclusively diagrammatic. Vesalius benefited from the artistic development of perspective which was introduced in the

15th century. Of the original 277 woodcut blocks, from which his drawings were printed, 227 survived a series of disappearances and rediscoveries and were used again and again in a variety of publications, the latest of which being issued in 1934!

William Harvey and De Motu Cordis

Vesalius taught anatomy in the School of Medicine at the University of Padua in Italy. This was one of Europe's great institutions of higher learning, and its anatomists were to extend Vesalius's efforts in succeeding generations. In 1599, Heironymus Fabricius of Aquapendente was professor of anatomy at Padua when a young Englishman, William Harvey, who had earned his B.A. at Cambridge, enrolled in the medical school. Fabricius had observed little flaplike structures in the veins of the leg. He drew excellent pictures which showed that the flaps point in one direction and would prevent the flow of blood in the opposite direction. Oddly enough (from our point of view) he didn't make much of an issue of his finding. His student, Harvey, would.

In 1602 Harvey returned to London and opened his medical practice. By 1615 he had received an appointment as professor of anatomy at the College of Physicians and Surgeons. From his days in Padua Harvey had been increasingly convinced that blood must somehow be conserved; it was, to him, inconceivable that it simply was endlessly produced only to be consumed. What had been no problem for Galen became a sticking point for Harvey.

As mentioned previously, his teacher, Fabricius, had discovered flaplike structures in the veins. These flaps were constructed so that fluid could pass through in one direction but would close if it attempted to flow in the opposite direction. The flaps could be interpreted as valves. Valves were in widespread use in all sorts of plumbing devices, and their purpose was clearly understood. This evidence of a one-way flow undoubtedly contributed to Harvey's thinking about the movement of blood, but the most impressive part of his thought was less concerned with anatomical detail than it was with a conceptual problem. Harvey's method of dealing with his concern about the amount of blood involved in the

Galenic explanation is an entirely new approach to questioning. Harvey estimated the volume of blood that the left side of the heart contains when it is full and found it to be 3 ounces. He estimated that the heart beats 33 times per minute. This is less than half of its actual rate so he was being extremely conservative. He continued his conservatism and assumed that only 1/4 of the 3 ounces (3/4 of an ounce) will be expelled from the heart with each beat. We have no way of knowing why Harvey underestimated the amount of blood, but if we use his figures and do the arithmetic we calculate that every minute the heart will expel almost 25 ounces of blood. In 1 hour close to 1500 ounces will have been expelled, about 90 pounds of blood. Harvey must have been amazed at these figures because he reduced them somewhat and claimed that in 1 hour the heart would pump only 60 pounds of blood. Even this amount is much more blood than is contained in the entire body. If we use his figures we arrive at 1440 pounds of blood leaving the heart every day!

Nobody before Harvey had taken the approach of calculating how much blood would be involved in the Galenic processes of *pneuma* formation. It is clear that Harvey is unwilling to have prodigious amounts of blood disappear and be reformed and his method of questioning is designed to stress what he perceives to be a serious problem in the Galenic description of *pneuma* formation. Harvey is not arguing against the formation of the *pneuma*, but the world of London and Padua in the 17th century is not one in which half a ton of blood appearing and disappearing daily can be accepted without question. In other words, the Galenic explanation is no longer completely *satisfying*. The possibility that blood is circulating, going round and round and not being used up and reformed, fits more comfortably with Harvey's values and sense of the way the world must operate. His quantitative observations, measurements, and calculations serve to provide evidence that his beliefs are correct. In other words, his convictions came first, and then he designed a way to ask the question in a manner which would force Nature to yield a response indicating the validity of his views.

"In other words, his convictions came first, and then he designed a way to ask the question in a manner which would force Nature to yield a response indicating the validity of his views."

Harvey published his theory of the circulation of the blood in 1628. The book, *Exercitatio Anatomica de Motu Cordis et Sanguinis in Animalibus* (Treatise on the

Anatomy of the Motion of the Heart and Blood of Animals) is familiarly known as *De Motu Cordis*. With justification, it is often cited as being the first truly modern approach to the study of anatomy and physiology. But don't get too carried away with Harvey's accomplishments. He left some gaps. One gap is structural, and the other is functional. What is informative about these gaps is the way he explained them and our reaction to his explanations.

A major problem with the circulation explanation is that there is a missing anatomical piece. The principal arteries are quite large, and their early branches are clearly visible. But as the arteries branch again and again within the organs, they become so small as to eventually become invisible. They simply disappear. Veins emerging from the various organs can also be seen, and if one attempts to trace their branches back into the organs they too disappear from sight without ever making contact with the incoming arteries. The critical *connection* between the arteries and the veins just isn't there! The circulatory theory requires a way for blood to actually make the circuit, but Harvey could not find the linking connection.

You may recall that Galen had a similar problem with the supposed pores in the *septum*. His theory required pores so he created pores. He solved his problem by assuming that they were too small to be seen. Harvey's explanation was more daring. He maintained that the missing connections were *incorporeal*, that is, without physical existence. Even more, he announced, it was their very lack of physical substance that enabled the maintenance of life. The pneuma could reach all parts of the body more effectively.

For us, today, there are two unsatisfying qualities to Harvey's explanation. We don't like the missing piece of anatomy, and we are uneasy about this pneuma business. Why is it that for thousands of years science could accept spiritous forces and incorporeal anatomy? This is the blending of fact and speculation that causes us to be uneasy about older scientific explanations. I mentioned our acceptance of invisible atomic particles (electrons and protons, for example) as explanations for the behavior of chemical substances. Each age has both factual evidence and speculative thought which it respects. Our task, in this book, is to trace the path by which respect has moved from *pneuma* to nucleic acids.

The microscope came into common use later in the 17th century, and the missing physical connections, the capillaries, were finally observed. The plumbing really makes a complete circuit. Harvey's claim that the final connection was incorporeal and that this was an important attribute for life was technically wrong but turns out to have a peculiar relationship with the truth. Capillaries actually have extremely *thin* walls which enable the passage of nutrition and respiratory gases without which life is indeed impossible.

"One of the great disservices we do is to use our present state of understanding to look back in time and selectively identify those whom we can turn into heroes by putting our words into their mouths. Even worse, we can identify as fools those whose thoughts, entirely valid in their time and place, have proven to be unsuitable in ours."

But we must not transpose our knowledge back to Harvey's time. He said what he did for *his* reasons. Within the structure of his model of reality his thoughts were coherent and valid. One of the great disservices we do is to use our present state of understanding to look back in time and selectively identify those whom we can turn into heroes by putting our words into their mouths. Even worse, we can identify as fools those whose thoughts, entirely valid in their time and place, have proven to be unsuitable in ours.

Natural Philosophy

The third contribution to modern biological thought was Greek **natural philosophy**. The nature of the physical world was perhaps the earliest target, but the philosophers had formulated advanced concepts as to the nature of life as long ago as 700 B.C. The preserved writings of the Greeks tend to surprise us because some of their thinking sounds quite "modern." Unlike much of the content of natural history, these speculations are not about individual organisms but instead are investigations into the nature of the laws under which life operated.

The Greek natural philosophers attempted to grasp the fundamental nature of the living state. They knew it differed from material which was not alive — earth, water, air, fire, for example. They devised systems of thought about matter, the most famous of which, the atomic theory of Democritus (470-380 B.C.), assumed that the universe was made of invisibly small particles (atoms) which had a variety of shapes and which could combine with one another to form all of the visible things of the world. Living things were also made of various atoms and gained their characteristics from specific combinations. Keep this thought in mind. This insight as to the variety of ways in which atoms might combine, thus yielding a variety of patterns, will surface again and again. It is a very enduring idea.

A dead body possesses all its atoms; if weighed a moment after death it will prove not to have lost any substance. One explanation for death is that some *weightless* essence, a vital spirit, has left the body. This was the most favored explanation for most of human history and still has tremendous appeal. But it is also possible that what may have been lost is the critical *pattern* of atomic arrangement. If you think about these two alternatives you may see that it probably is not possible to decide between them.

The Search for Order and Harmony: Greek Patterns of Meaning

The Greek natural philosophers were impelled to create patterns of order and to search for meaningful relationships in Nature. From the very early assertion that the world consisted of four fundamental entities — earth, air, fire, and water — they extended their logic and assigned to these the four qualities, dry, wet, cold, and heat, respectively. Living things clearly possessed these qualities in varying degrees. Obsessed with the conviction that what is correct and true and healthy is based on order and harmony it was a very natural step to perceive illness as being a departure from a balance of these qualities.

Empedocles (504-433 B.C.) made a significant effort to merge natural philosophy with practical medicine. He participated in the creation of what was to become the practice of "rational" Greek medicine. Sometimes this is spoken of as a belief in natural causes as contrasted with an earlier practice which was dependent upon "sacred" causes. Building upon the concept of the four elements together with their properties, Empedocles proposed that the qualities dry, cold, hot, and wet were represented in the body by four "humors" — blood, black bile, phlegm, and yellow bile. Further, he proposed that blood was produced by the heart, black bile by the spleen, phlegm by the brain, and yellow bile by the liver. It was the proper interaction of the four humors which gave a body its health.

In this recounting of Greek thought there is the danger that it may blur into a meaningless jumble of ideas and names and the reader may become concerned that the details are intended to be memorized. The point of this discussion is to emphasize the Greek passion for constructing patterns of meaning. The natural philosophers were probing for ways to build mental frameworks upon which to suspend the factual pieces of reality which were beyond argument. Blood and phlegm exist as certainly as fire and air do. Is there a pattern of relationship?

Hippocrates (460-377 B.C.), used the humoral theories of the natural philosophers as the basis of his medical practice. If a balance of humoral factors signaled good health then the loss of that balance explained illness. The physician's role was to determine what had become unbalanced and to restore the equilibrium of health. Since the distant days beyond memory the human race had used herbs, roots, minerals, oils, leaves, horns, and other substances in its efforts to heal and forestall death. Hippocrates was well aware of the effects of various substances and procedures. What he was searching for was a way to *explain* what occurred.

"The Greek physicians were attempting a very daunting task. They were drawing all of the observable world into a single unifying explanation. They were building connections between the things they could see and attempting to find hidden meanings beneath the surface of the visible creation."

The Greek physicians were attempting a very daunting task. They were drawing all of the observable world into a single unifying explanation. They were building connections between the things they could see and attempting to find hidden meanings beneath the surface of the visible creation. This is exactly what science does today, and it is our task to discover how and why the way of doing science changed.

Medieval Scholasticism and the Renaissance View of Life

In the effort to make the history of scientific thought meaningful we cannot just select some date and pick the story up at that arbitrary point. It isn't possible to start in 1953, for example, and give the explanation most in favor that year without immediately raising the question as to why people thought that would be a good explanation. What was the previous explanation? What was unsatisfactory about the previous one? We are describing what has been an ongoing process, and it is only by this procedure that we can reach a personal decision as which views of life, if any, are meaningful to us. I don't think it will be "giving away the ending" to say that for some readers, today's explanation will be less satisfying than some earlier views.

The Middle Ages are typically dated from A.D. 476 to 1453, the start of the Renaissance. St. Thomas Aquinas lived from 1225 to 1274 and his thought is indicative of the way in which life was viewed during the latter part of the Middle Ages. Thomistic thought was reliant upon and interpretive of the philosophical tradition of ancient Greece. There were two conflicting views of reality argued by the Greek philosophers. Plato's world came into existence as the concrete expression of the Creator's images. For Plato, the things we see, be they stars or starfish, are expressions of *forms* or *ideals*. The Creator-Craftsman imposed various forms upon matter to create the universe we see. In this way Plato explained the species of living organisms. Each "kind" of animal and plant preexisted as an ideal before it came into physical being. The clay, as it takes on form, reveals to us the potter's *intention* as it becomes a vase or a bowl. So if we see horses or camels we know that each of these is a distinct "kind" of organism because each is a physical manifestation of the Craftsman's intention. For Plato there was a clearly defined beginning of Creation and when the Creator was finished forming all of the intended creatures, creation ceased. The Platonic world is a finished place. It is enduring and unchanging.

"To jump ahead in our story, Charles Darwin was not the first person to suggest that creatures can evolve. That idea was thousands of years old before Darwin took his famous trip on The Beagle."

Aristotle modified Plato's thinking. He injected the idea that the *forms* or *ideals* don't exist in some separate place (like the mind of the Creator) but in the creatures themselves. For Aristotle each ideal is constantly renewed and recycled as the parent organisms die and their offspring become the manifestation of the ideal. Aristotle's world was being *constantly* created. He did not believe that the universe came into being at some time in the past and that since then it has been static. This is an important distinction. In it we will find the seeds of the idea that creatures can change over time. To jump ahead in our story, Charles Darwin was *not* the first person to suggest that creatures can evolve. That idea was thousands of years old before Darwin took his famous trip on *The Beagle*.

Not all Greek philosophers liked the concept of the ideal. Democritus's atomic theory did not assume that the atoms formed themselves into beings based on a set of ideal forms. Quite the contrary. The so-called *atomists* maintained that nothing existed except the bits and fragments swirling in a mindless and formless void. To the atomists there was no *intention* in creation. The most fully descriptive statement we have expressing this philosophy is transmitted by the Roman philosopher-poet, Lucretius. Writing about 100 B.C., in *De Rerum Natura* (On the Nature of Things), Lucretius said that the shapes of the particles and their various motions are the only forces in the universe. Matter and motion is all there is, said this message of pure mechanism. The organisms we see, ourselves included, are shaped by blindly occurring collisions of mindless atomic particles. This is a very sobering view of reality. It has no intention, no goals, no purpose. This view of life came to dominate scientific thinking, but we are getting ahead of the story.

For the 13th-century mind of Thomas Aquinas there was little doubt as to which account of creation squared best with Christian doctrine. Plato's concept of ideals in the mind of a Creator was much more attractive than atomic chaos. It was a Craftsman-God that St. Thomas saw operating throughout the first 6 days of creation. By using Platonic forms, this God created each species of animal and plant with specific intent. Aristotle's modification was also worked into the Thomistic explanation. Once God was finished creating the various species, He ordained that the creatures were to assist in the task of maintaining their kinds. Each organism contained, within itself, the form (some writers use the term, "soul") for its species and by the process of generation the creatures continued the creative act originated by the Craftsman. It was this combining of Greek thought with the biblical account of the creation in Genesis which was to have such a profound effect upon the minds of biological scholars with regard to the species question. We will encounter the Aristotelian/Thomistic perceptions throughout the history of the evolution debate which we will discuss in the following chapter.

By the time the Renaissance arrived many of the the

"This passion for seeing relationships was driven by the conviction that there was great wisdom to be had if one could perceive the deeper lying truths which were hinted at by the similitudes. The forces which shape stars and men were there, just below the surface, waiting to be grasped."

Greek views of life had become less distinct, more cluttered. The efforts of medieval scholars had added additional layers of observations and inferences. If a creature was shaped like a star what was to be inferred from this similarity? What was Nature's message to us when we found a plant root shaped like a human form?

This passion for seeing relationships was driven by the conviction that there was great wisdom to be had if one could perceive the deeper lying truths which were hinted at by the similitudes. The forces which shape stars and men were there, just below the surface, waiting to be grasped. Viewed from this perspective it becomes possible for us to understand how the Renaissance mind could associate a plant with eye-shaped markings with a cure for blindness. The scholars of this period were pursuing a line of thought which placed great value on visible manifestations of hidden relationships. Did children not resemble their parents? Wasn't this evidence of a hidden yet powerful linkage between the generations?

We've come to a point in this history of ideas where we have sufficient background to understand how the study of the living world was to become dramatically changed by changes in thinking in physics and astronomy which would affect not only views of life but political and social systems as well.

The 17th century was a time of turbulent ideas. Galileo (1564-1642) was to force the issue of an earth displaced from the center of the universe, and Descartes (1596-1650) was to disconnect his century from Aristotle's influence on all thinking about reality. Sir Isaac Newton (1642-1727) would devise a system to explain matter and motion which was to prove so powerful that its explanatory and predictive qualities have survived all but the most recent modifications of modern physics. The 17th century introduced the concept of natural law, a set of mechanistic rules by which apples fell and planets moved. The search for relationships would continue but in an entirely different manner. Visible similarities would no longer point to possible linkages. Instead, the immutable laws governing matter and motion would apply to all things, and the activities of all creation would be examined in the light of these governing controls.

In the following chapter we will see the influence of this newly freed and enlightened thought on the essential problems of life. Was it possible to explain the mysteries of life using the intellectual tools of mechanism?

CHAPTER THREE

THE GREAT CHAIN OF BEING

"[T]he number of true species in nature is fixed and limited and, as we may reasonably believe, constant and unchangeable from the first creation to the present day."

— John Ray

Aristotle, having studied the natural world more thoughtfully and comprehensively than anyone before him, is usually credited with having perceived of all of existence as a Great Chain of Being. We have previously mentioned Plato's ladder of life, but the Aristotelian view of the universe included not only organisms but every existent thing in a continuum of creation. From the gods and their world in the heavens, the sun and the moon, to the stars and the planets, seas, mountains, deserts, grains of sand, and the smallest of the creatures, Aristotle and those who followed his teachings arranged all things in a Great Chain of Being. This metaphor has a magnificence about it. We see a chain, its links shining and constant through eternity.

Metaphors have enormous power. A chain is linear; its links follow one another in a straight unbroken line. Each link has a location. Links do not change; they do not come and go. The image of a chain implies permanence, unity, and joined relationships.

Aristotle could have used a different metaphor. Why not a Great Web of Being? Or perhaps a Great Tree of Being? This chapter will explore the influence of metaphors upon the centuries of thought which follow their adoption.

Models and Metaphors for Origins

When I was a high school student of biology I was told that the earliest mention of **evolution** was by the Greek philosophers Thales (639-544 B.C.) and Anaximander (611-547 B.C.). I thought it was extraordinarily perceptive for them to have had such a "modern" idea. As I think of it now I wonder why I thought evolution was a recent concept? One of my problems was in separating *accounts of origins* from *explanations of species*. Thales and Anaximander were speculating as to how the world (we might say the universe) came into existence. I thought they had speculated as to how species had evolved. These are two different issues.

"As I think of it now I wonder why I thought evolution was a recent concept?"

There are two major alternative explanations as to how our world came into being. The first is that each of its parts was created separately as we see them now. Most creation accounts are of this type. The image that comes to my mind is the unpacking of a new chess set. There's the board and all the chess pieces — knights, pawns, bishops, king, and queen — all present and ready for play. All of the pieces have certain qualities, moves that each may make and rules that govern their play. The rules do not change. An interesting quality of the chess set image is that the nature of the board and the appearances of the pieces imply that there is some *purpose* or *intention* associated with the set.

The second explanatory alternative is that the world is the result of a process of change from some primeval starting condition. A metaphor for this is a new chemistry set. There are no rules. Each of the substances may be combined with any of the others. Rows of little containers of chemicals and a few test tubes don't provide any indication of the potential that the set possesses, and there isn't even a hint as to what the intended outcomes are. But a chemistry set will indeed produce outcomes even though none are evident when the box is first opened.

The belief that the origin of the world resulted from a process of successive changes is a very ancient idea. Thales, possibly influenced by the relatively large amount of water making up his coastal world on the Aegean Sea, suggested that water was the primeval element from which all other elements were derived. Anaximander modified this view and concluded that air and earth were also primeval and that along with powers of heat and cold, the water, air, and earth were continually being modified into the created things of the world; creation for Anaximander was still under way in a world of constant change. He went further and proposed that the first living things appeared in the water and that by processes of change in the mixtures of elements and forces these first creatures became land organisms. Human beings as well as all other creatures were derived from these aquatic predecessors according to Anaximander.

The biblical account of creation in Genesis is of the chess set variety. With two specifically stated exceptions, it informs us that each component of the world was *made* uniquely; they were not modified from something else. The only exceptions to this generalization are Adam and Eve. Adam was formed from the "dust of the ground," and Eve was formed from Adam's rib. We have been cautioned, by biblical scholars, that the Bible is neither a scientific nor a philosophical treatise and that we should not read it as such. Much of the argument concerning origins and species does, however, center upon biblical statements and their interpretation.

Cogito, Ergo Sum

Although the beginnings of modern attitudes toward the nature of life can be found in bits and pieces stretching back into antiquity, René Descartes (1596-1650) provided us with a reasonable place to begin. Cartesian thought has been viewed as the bridge between Scholasticism (with its roots in antiquity) and all of the philosophical thought which was to follow. The Cartesian position was that doubt clearly exists and authoritarian statements cannot displace it. So Descartes began with the existence of doubt and himself as a doubter. With confidence in mathematical treatment of universal mechanistic laws, a doubting person can extend knowledge and gradually replace doubt with comprehension. The Cartesian assertion —*I think, therefore I am* — is a departure from all previous efforts to establish the reality and nature of existence. Taking his own existence as the starting point, Descartes moved in successive steps to the consideration of all other doubted information.

"Taking his own existence as the starting point, Descartes moved in successive steps to the consideration of all other doubted information."

Cartesian and Newtonian Mechanism

Descartes stated a belief that God exists as the first cause from which all other causality flows. In considering how God might have acted in creation, he suggested that the present earth *could* have been brought into existence, had God wished to do so, by a series of gradual changes from a primitive state. He described a possible method of formation of the planets from coalescing clouds of matter; since he was convinced that the earth is a planet circling the sun, this possible method could account for our earth's origin. Descartes, in keeping with the practice of his time, stated that God *actually* created the world in the manner described in Genesis, but by indicating a *process* using natural mechanisms he established an intellectual position from which science could phrase questions freed from the authority of antiquity.

Isaac Newton (1642-1727) posed questions that Nature could answer directly and with revealing clarity. He saw that it was the *change* in the quantity of a body's motion that accurately measured the force acting upon that body. We assume, today, that it has always been understood that if a moving object speeds up or slows down the increase or decrease in its motion is a reflection of the forces acting upon it. It was Newton who pointed out that *measuring* change was the most effective way to reveal the qualities of Nature's hidden forces by quantifying the effects of these forces. Measured change revealed an entire world of reality that lay beneath the visible surface. This approach allowed Newton to correct prior errors in physics and his Laws of Motion enabled him and those who followed to bring coherent unity into what had been a jumble of disconnected problems. A unified under-

"He saw that it was the change *in the quantity of a body's motion that accurately measured the force acting upon that body. "*

standing of gravity permitted the motion of planets and the leap of a deer to be studied by using the guiding influence of one universal set of laws. The Web of Reason could be spun connecting the vast array of events in the heavens and on the earth. Testable mechanisms rather than possible but untestable powers were to be the targets of scientific investigation.

"The Web of Reason could be spun connecting the vast array of events in the heavens and on the earth. Testable mechanisms rather than possible but untestable powers were to be the targets of scientific investigation."

Mechanism in Origins and Species Formation

It is impossible to overstate the dominant influence of the **mechanistic attitude** on the way science would approach all aspects of life following the introduction of Newtonian physics. It is one thing to call up the *image* of a pump in discussing the heart, and quite another to be convinced that the heart *is indeed* a pump in every physical sense. We speak of a "broken" heart not as a failure in circulatory capability but as the indicator of a failed love. To have "heart" is to be courageous. Once the heart is seen as "only" a pump these other meanings become quaint vestiges of an earlier set of beliefs. The mechanistic approach to the study of life began with a vengeance in the 18th century. Some of these efforts would prove to be successful. Others were doomed to failure because of an underestimation of the difficulties involved.

Buffon and the Internal Mold

The imaginative use of the mechanistic approach is well illustrated by the work of the French naturalist, George Louis Leclerc, comte de Buffon (1707-1788). In a monumental 44-volume compendium, *Histoire naturelle*, written between 1749 and 1804, Buffon attempted to explain life's attributes in the light of Newton's laws. One of the truly mysterious aspects of life is the way in which food becomes transformed into the substance of the living organism. Buffon attempted to explain this conversion mechanistically by beginning with the proposition that all organisms are composed of organic molecules which were formed from primeval atoms. This is obviously an idea influenced by Democritus by way of Lucretius. Buffon reasoned that what makes one species different from another is the *pattern* of the organic molecules. There must be a dog pattern, a cat pattern, a rose pattern, etc. Each creature's pattern, suggested Buffon, functions

"Buffon reasoned that what makes one species differ- ent from another is the pattern *of the organic molecules."*

as an *internal mold*. Just as an external mold shapes the liquid plaster poured into it, so does the internal mold function to organize the organic molecules contained in the food.

This mechanistic analogy of Buffon's may appear either convincing or not depending upon one's need for detail. If the general idea of a mold imposing form upon matter is satisfying to you then the concept is at least a possibil- ity. But if you demand to see the mold and examine exactly how it forces the organic molecules into the desired pattern then Buffon leaves you unconvinced. Let's give him a little help.

Instead of an *internal mold* organizing the food mol- ecules, let's use the word *blueprint*. Each species is built upon a *plan* using a blueprint specific for its structure. Does this sound more convincing? No? Let's make one more adjustment. Let's substitute for the word "blue- print" the words *"genetic code."* Now the food molecules are organized, according to a genetic code, into the pattern of organization of the species involved. Are we coming closer to making sense?

The Use of Models in Creating Explanations

It takes a little mental stretching to properly examine ideas. Buffon's *internal mold* is not really a *thing* to be seen and touched as much as it is a *capability*. Realizing that food must have some guidance in the process of becoming flesh, Buffon examined the world he knew and understood for a possible *mechanism* which would have such a capability. He settled upon the mold as a model of the kind of mechanism with the necessary capabilities.

"Buffon's internal mold *is not really a thing to be seen and touched as much as it is a* capability. *"*

Is a model, whether a physical one (like a mold or a blueprint) or an intellectual one (such as a genetic code), useful in explaining Nature? In thinking about all of the different species distributed on the earth, Buffon relied upon the mechanistic Newtonian vision of reality. There were two very strong forces acting upon all matter, including the organic molecules Buffon considered to be the material out of which life was constructed. Newtonian **gravity** was a universal and unchanging force. Buffon argued that gravity would attract all organic molecules to one another with a constant force. Heat, however, was observed to create expansion which indicated that molecules were being pushed apart. Heat was a very variable force; some places on earth were much warmer than others. Buffon had a model with which to explain diversity.

Buffon felt that the early earth had been much hotter than at present. He arrived at this conclusion by applying reason to the mechanistic model. If the earth had been formed from the coalescence of a particle cloud then the force of gravity must have been operative in pulling all the particles toward one another to a center. What force had pushed the particles apart in the first place? Heat. So in the process of earth's acquiring its present form its temperature must have declined so as to permit gravity to win out in the balance of forces. Remember that Newtonian gravity remains constant so it cannot be that the force of gravity increased. It must have been that the heat decreased.

As the earth cooled, organic molecules coalesced from primitive atoms. Depending upon the amount of heat available, different kinds of organic molecules in varying abundances would have formed. As further coalescence occurred, these molecules assumed various patterns (internal molds). It was these patterned assemblages of molecules that gave rise to the first living things according to Buffon's explanation.

Different kinds of animals appear in various geographical distributions. Buffon explained why we see giraffes in Africa and not in South America as the result of differing conditions at the time that patterns were being established. Not only were there differing amounts of heat, he proposed, but also differing amounts and kinds of organic molecules. All of us who were raised in the Great Plains states expect dirt to be brown or black. Our first trip to the mountains of Colorado revealed why the Spanish explorers named the region as they did. The erosion of its red sandstone rock created an entirely different earth color. Buffon was aware of this kind of geographical complexity and saw in it the basis for animal and plant diversity.

Starting with confidence in Newtonian physics and mechanistic explanations together with his model of an *internal mold*, Buffon created a coherent vision of the early earth and its subsequent history. The diversity of life could be rationally explained. Considering the information available to him, this was an extremely impressive accomplishment.

Carl Linné and God's Secret Cabinet

An accomplishment of an entirely different sort was made by a Swede who was born in the same year as Buffon, Carl Linné (1707-1778). Following a scholarly tradition he latinized his name to Carolus Linnaeus. Linnaeus had a somewhat different agenda. He was not interested in theories of the earth's formation but saw himself as continuing the great natural history tradition of classification. Placing the world's organisms in meaningful arrangements, he felt, would reveal the underlying rules governing similarity and difference. To this day we use the **binomial** (two-name) system of nomenclature devised by Linnaeus. Every time we say *Homo sapiens* in identifying our species we acknowledge his orderly system of naming creatures.

The perception that Linnaeus was an eccentric figure wandering through the countryside collecting and naming organisms needs to be dispelled in order to understand what he was about and the position he held in his own time. Linnaeus studied medicine which, in his day, was heavily influenced by botany owing to the medicinal qualities of plants. Sweden, at the time of Linnaeus's birth, was an imperial power, but in the Great Northern War fought against Russia, Poland, Denmark, and Prussia it lost its Baltic empire. The nation was bankrupt and sought to regain solvency by identifying its internal resources. To this end expeditions were mounted to identify natural resources, and it was this effort that provided an outlet for the naturalistic talents of the young Linnaeus. Although he had essentially completed his medical studies and had written the required thesis, he had not actually been awarded his doctorate. An opportunity to continue his studies in Holland arose, and he took it.

Holland was the intellectual center of Europe. It also had worldwide colonial and trading interests. Its merchants received, from the Indies, for example, botanical and geological specimens in enormous numbers. All of this had to be cataloged and evaluated as to commercial potential. Linnaeus's skills at identification and classification were further developed under Herman Boerhaave, the most distinguished scientist of the time. Under Boerhaave, Linnaeus continued his study of chemistry and botany.

It was critical for his financial stability that Linnaeus be awarded his doctoral degree. In Holland was the University of Harderwijk, an institution which was an acknowledged diploma mill. Taking his thesis (on the cause and cure of intermittent fever) to Harderwijk, Linnaeus presented it, was examined upon it, and had his degree awarded all within the space of one week. His degree in hand, he was prepared to spend his life in the scholarly investigation of the organisms of his world.

We identify Linnaeus as the person whose system for naming creatures is the one in use throughout the scientific world, but more important than his system of nomenclature was his vision of classification (taxonomy). A **taxon** is a group of organisms in a formal system of classification. Linnaeus followed in the Aristotelian tradition of classifying organisms **hierarchically**. Taxa (the plural form) are arranged in increasingly inclusive groupings, as in a series of nested boxes with the largest box eventually including all of the smaller ones. Most of us know that humans are all members of the **species** *sapiens* within the **genus** *Homo.* There are other species in our genus. For example, *Homo habilis* appeared on earth about 2 million years ago and became extinct about half a million years later. By successively grouping related **taxa,** Linnaeus's system creates larger and larger groupings. With the species as the smallest taxon, the successively larger groups are called genus, family, order, class, subphylum, phylum, and kingdom. (There are several intermediate groups — e.g., subspecies, superclass — but we are concerned with the pattern, not the details.)

It is obvious that the essential challenge in a hierarchical taxonomy is deciding upon the *basis* for the groupings. Which features are critical in deciding which species are sufficiently similar to group in the next highest taxon, the genus? Which genera (plural of genus) are to be grouped

"It is obvious that the essential challenge in a hierarchical taxonomy is deciding upon the basis for the groupings. Which features are critical in deciding which species are sufficiently similar to group in the next highest taxon...?"

into families, which families into orders, and so forth? This calls for a degree of assurance which most do not possess. It is fascinating, in this regard, to read in Linnaeus's autobiography his description of himself:

> God has suffered him to peep into His secret cabinet. God has suffered him to see more of his created worke than any mortal before. God has endowed him with the greatest insight into natural knowledge, greater than any has ever gained. The Lord has been with him, whithersoever he has gone, and has exterminated all his enemies, and has made him a great man....

The reason for quoting this self-evaluation is not to present Linnaeus as an overly pompous person but to establish the basis for the nature of the taxonomic controversy which raged during the 18th and early 19th centuries. The "enemies" who have been "exterminated" were those naturalists who disagreed with Linnaeus as to the proper structure of a taxonomic system. Having identified characteristics as crucial, naturalists would rearrange their classifications to reflect *their* vision of the living world. Organisms have all sorts of structures, both external and internal, and it is obvious that there will be differences of opinion as to which are to be considered the real indicators of relationship.

The Basis of Relationship I.
Homology in a Static World

Let's recall the two analogies I used in considering the alternative methods of creation, the chess set and the chemistry set. Linnaeus favored the chess set model, a world whose species had all been created at some distant time in the past. No new pieces had been added. These pieces, the organisms, were found distributed about on the planet, and it was the task of the naturalist to seek the identifying characteristics which would permit orderly and meaningful grouping. Exactly which characteristics would reveal the relationships?

The problem was in the identification of those structural features which were indicative of fundamental similarity. Just what was *fundamental*? Most of us would reject color or overall shape as a fundamental quality. Being green doesn't indicate a fundamental relationship between frogs and spinach. Similarly, worms and snakes differ too much in their internal anatomy to get more than a momentary consideration as relatable. What impresses us more than anything else is a structural quality that we see as an **organizational pattern**.

Since our earliest exposure to skeletons we have been struck by the *pattern* of the bones in a human arm and hand as being hauntingly repeated in the forelimbs of apes, cats, bats, and the wings of birds. We've seen skulls whose shapes may be distortions of our own, but the same bones appear in anticipated places. There is something of an underlying theme in these perceptions. It does not matter that a bone may be longer than another or more curved than flat, it is the *arrangement* which demands acceptance. Biologists use the term **homologous** in speaking of anatomical structures which communicate this identity of pattern. The arm, wrist, hand, and finger bones of a human are said to be homologous with the wing bones of the bat because they are all in their anticipated locations, demonstrating a unity of pattern. We observe the dramatically elongated nature of the fingers in the bat as they create the skeletal support for the membranous wing. We find the same pattern in the flipper of a whale, where the arm bones are dramatically shortened but the basic arrangement is unmistakable.

"Since our earliest exposure to skeletons we have been struck by the pattern *of the bones in a human arm and hand as being hauntingly repeated in the forelimbs of apes, cats, bats, and the wings of birds."*

From the viewpoint of **homology** all organisms can be seen as variants of basic organizational plans. All mammals can be seen as variants of a fundamental mammalian model. Not only are all catlike mammals clearly relatable, but the cat family skeleton is echoed in the skeletons of horses, apes, and humans. It is as if the Great Chain of Being had sections where all of its links were slight variants of some basic theme. The mammalian section of the chain would consist of clearly relatable variants: dogs, cats, apes, and humans. Further along the chain would be the reptile section where a reptilian model had been expressed in numerous variations. What is striking is that the reptilian model and the mammalian model had very obvious homologies. This relatability extended to the fish section of the chain and the bird section as well. In other words, the detectable pattern extended in all directions, becoming progressively less distinct with distance.

"From the viewpoint of **homology** *all organisms can be seen as variants of basic organizational plans."*

The entirety of the Chain of Being could be thought of as consisting of these sections, each section being a basic design region within which a homologous pattern united all members of the section. From any starting point, one could move to neighboring reaches of the chain where even though faint, the pattern was detectable. At very great distances one would encounter design patterns which had no relatability to the starting point.

Let's consider another metaphor for homologous patterns. Imagine a pond into which a handful of pebbles has been thrown. Each stone creates the center of a homologous pattern from which spreads a ripple extending that particular design. There is the vertebrate center with its recognizable ripple extending to include fish, reptiles, birds, and mammals. Some stones produce ripples which encompass the homologous patterns of wormlike creatures; others the radial pattern of starfish and sea urchins; still other stones spread the ripples of microscopically small creatures.

I have presented two metaphors that are intentionally incompatible. The Great Chain of Being is, first and foremost, enduringly linear. Its links are permanent; the various sections were created and forged into position never to be moved. The sequence stretches from end to end with each individual link unchanging both as to its makeup and its location.

The metaphor of many stones, each creating its own center, is neither linear nor enduring. The centers are scattered. The ripples spread over time. The pond has no obvious beginning or end points. If one were to have looked at the pattern of ripples yesterday it would not have been the same as it is this morning. By tomorrow the pattern will have changed.

The selection of a metaphor forces us to make decisions as to just what we think the metaphor represents. I chose to place these incompatible alternatives in opposition to one another to introduce a second perspective with regard to homologous patterns.

The Basis of Relationship II.
Homology in a Changing World

Where should one place fossils in the Great Chain of Being? We can imagine the first time this challenge presented itself to the early naturalists. Skeletons of familiar creatures pose no real problem: one's living dog and a dog skeleton are obviously one and the same link in the chain. But how about a fossil shell for which no living creature's shell can be found?

There are two ways to deal with fossils. In *appearance*, there is no doubt that this rocklike thing has a relationship to life; fossils look so much like entire fish or bones or leaves or even feathers that it is impossible to believe that they were ever just ignored as being oddly shaped but quite ordinary rocks. No, these particular objects have to be explained. One explanation is that some set of nonliving forces created them to *appear* like organisms or their parts. We've all encountered cliffs which have eroded to the point where the profile of a person can be observed; such a profile is just a physical phenomenon. The problem for this approach is the uncanny precision of the fossil. It's just too perfect to be shrugged off as an accidental accumulation of haphazard knocks and flakings which somehow produced a grinning skull. It might well be evidence of sinister forces which are *not* natural. Medieval and Renaissance speculation concerning fossils frequently made the assumption that they were manifestations of evil intent.

The alternative explanation is that fossils really *are* what they appear to be, the remains of long dead creatures. If this approach is adopted then the fossils must take their places in the classification charts and the Great Chain of Being. Just because a "kind" of organism is no longer present doesn't disqualify it from inclusion in the list of created things. A favorite explanation was that the Great Flood had eliminated scores of "kinds" in spite of Noah's efforts to gather up representatives of all living creatures.

We've already mentioned the classical Greek suggestion that the first and simplest living things came into being by the transformation of inert matter and that further transformations of the simplest creatures resulted in the appearance of all of the species observed populating the present earth and its seas and skies. The idea that some species were unsuccessful and perished while others flourished was never doubted by many medieval and Renaissance naturalists.

There was a conviction that some of the existent species had been formed by the interbreeding of parent species. The creation of hybrids fascinated the medieval mind. In the imaginatively illustrated books called bestiaries, portrayals of the fanciful offspring of the sexual union of dissimilar animals and plants gave evidence of the conviction that hybridization did indeed occur. But it wasn't until the 18th century that the ancient idea that *all* species came into existence by the transformation of simpler ancestral organisms was seriously considered. This concept, **transformation,** was a direct contradiction of the Platonic/Aristotelian principle which held that each kind of creature was modeled on a unique ideal. It was equally contradictory of the biblical description of Genesis. But if the question as to the *cause* of homology in the living world was phrased within the context of transformation an entirely new view of life emerged.

*"This concept, **transformation,** was a direct contradiction of the Platonic/Aristotelian principle which held that each kind of creature was modeled on a unique ideal. It was equally contradictory of the biblical description of Genesis."*

If all mammals were descended from some primitive ancestor then they all would have inherited their basic skeletal pattern from this source. Variations obviously had arisen over time, but within the range of variability was embedded the basic design to be revealed by the homologous structures. And if the mammals all had a common ancestor, how was one to view the obvious homologies between mammalian and reptilian skeletons? Could it be that there was a common descent linking mammals and reptiles? The birds and amphibians and the fish had homologous connections with one another and with the reptiles and mammals. Were all of these vertebrate creatures members of a family by descent? Was the existence of homology indicative of a series of transformations which had produced the present living world from a parent stock of ancient creatures?

This was indeed a most serious challenge to the Platonic view of a population of ideal organisms. Rather than individually conceived and uniquely created, each kind of creature was actually a descendant modification of some unknown originator. This explanation for homologous patterning depended upon relationship and change.

It was dynamic, unpredictable, and very different from the majestic intention of a Platonic universe of eternal ideals.

CHAPTER FOUR

HISTORICAL CHANGE: AN 18TH-CENTURY CONVICTION

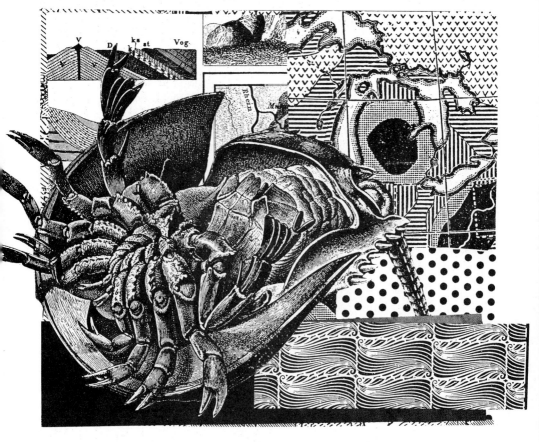

"An Organism is merely a transition, a stage between what was and what will be."

— François Jacob

The possibility that living organisms were modifications of prior creatures was only one aspect within a much broader perception which was to become a major force in the 18th century. This intellectual position was that **historical change** was the only satisfactory way to account for both the physical world and its living inhabitants.

According to this view, everything had a history of successive changes, and it was only by examining the present in terms of its past history that a rationally satisfying explanation could be developed.

Keep in mind that Newtonian physics and the persuasive power of mechanistic thought dominated 18th-century science. Natural law was raised to the level of belief previously reserved only for religious truths. In possibly unexpected areas of study the attractiveness of the historic change explanation was evidenced. A particularly informative example was the study of language. Eighteenth-century philologists found in such diverse languages as Sanskrit, Greek, and Latin, structural patterns which can only be described as homologous. The languages were all patterned upon a common theme. Just as bat wings and whale flippers don't look much alike at first glance, neither do these languages. Underlying the surface differences, however, were basic similarities. Could it be that the history of these languages, if traced backward, would lead to a common source? Viewed from this perspective, previously confusing qualities of language were satisfactorily explained.

Another subject area, geology, was beginning to give rise to historic attitudes. The various layers of sedimentary rocks were, in the 18th century, found to contain typical fossils. It did not matter that a limestone layer was seen in two locations 100 miles apart. If the layer at one point contained a distinctive collection of fossils these same fossils would be found 100 miles away. There were relatable qualities even to layers of rocks. When an English geologist wrote of his findings to a French colleague, the scientists learned that the same rock layer appeared on both sides of the English Channel. The unique population of fossils identified the layer like a fingerprint. Was it possible that the English and French rocks have had a common history?

"It is one thing to suggest that something has occurred and another to provide a satisfactory explanation as to how it occurred."

It was in this 18th-century intellectual climate that the first serious considerations of various *mechanisms* for the transformation of animals and plants were published. This point requires a reminder that although the ancient Greeks had believed in the descent and transformation of organisms from primitive ancestors, they had not speculated as to the way the transformation had actually occurred. It is one thing to suggest that something *has* occurred and another to provide a satisfactory explanation as to *how* it occurred.

Lamarck and Unlimited Mutability

The most influential statement as to a possible transformation *mechanism* was that of the French naturalist Jean Baptiste Pierre Antoine de Monet, chevalier de Lamarck (1744-1829). This man has been the subject of enormous ridicule and, only recently, a proper degree of admiration. We will encounter his name as identifying a school of thought, **Lamarckism**, in several future discussions.

Like Buffon before him, Lamarck was a Newtonian mechanist. He accepted the ability of the powers of gravity and heat to form organic molecules but he went considerably further in proposing how much change a living thing could undergo. For Lamarck, the effects of gravity and heat were only the beginning of the mutability, the changes, which life had undergone in the past and would undergo in the future.

In 1800, Lamarck tentatively presented his thoughts in his *Philosophie zoologique*; these thoughts were to be enlarged upon over a period of 15 years. The basic message was a formalization of the generally accepted belief that body parts which were used became larger and more fully developed while those which were not put to use decreased both in size and strength. This is such a common-sense reality that it is difficult not to accept it. The increased musculature of a blacksmith, the thickened skin on the palms of woodcutter, and even the specifically localized callus formation at the tips of a guitarist's fingers point to a use and disuse mechanism. The term "use" could be logically extended to a variety of involvements such those of the skin, which darkens when exposed to the sun while the covered regions of the body remain pale. Who has not observed the effect of disuse when, following a prolonged period of illness, the limbs have withered and the body's overall musculature is reduced and weakened?

*"Lamarck proposed that it is the **use** or **disuse** of body parts which provides the mechanism for their attainment of size and strength and, further, that once a part has acquired a change due to its activity, that **acquired characteristic** can be **inherited** by the offspring."*

Lamarck proposed that it is the **use** or **disuse** of body parts which provides the mechanism for their attainment of size and strength and, further, that once a part has acquired a change due to its activity, that **acquired characteristic** can be **inherited** by the offspring.

This "inheritance of acquired characteristics" was also a

commonly held belief. Returning to the blacksmith, weren't his son and grandson, who also worked in the shop, men of above average size and strength? Didn't both the coal miner and his sons, who shoveled at his side in the mine, all have calloused hands?

Lamarck suggested that the mechanism behind the changes in organisms was a life force which sought full expression of the potential possessed by the creatures. The giraffe's elongated neck was testimony not only to to generation after generation of stretching to acquire the leaves just beyond reach but also to the presence of an upward striving within all living things. It was this latter power, claimed Lamarck, which explained the *scala naturae,* the clearly evident progression from simpler to complex forms of life.

Lamarck's professional associates were suspicious of his upwardly striving force since it sounded like a return to a dependence upon mystical qualities rather than Newtonian mechanism. As to the inheritance of characteristics acquired through use and disuse, there was a mixed reception. Although the concept made rough sense, it wasn't clear that all, or for that matter even most, improvements were in the direction of increase. While this explanation worked for such obvious cases as the length of the giraffe's neck and the anteater's tongue, it seemed too crude a process to explain the incredibly fine detail which organisms displayed. For example, the three small bones (ossicles) of the middle ear are subjected to constant use throughout the lifetime of creatures in this noisy world. Should they not have become progressively more massive? There were too many deviations from the simple rules of use and disuse that would have to be imagined to adequately explain the appearances of organisms.

"It was rapidly becoming a major tenet of naturalistic thought that far from being immutable and unchanging, life was in a slow but continual state of change."

But there is no doubt that Lamarck's claim for unlimited mutability on the part of living things was greeted with enthusiasm. It was rapidly becoming a major tenet of naturalistic thought that far from being immutable and unchanging, life was in a slow but continual state of change.

One of the major influences on this shift in attitude was the geological speculation previously mentioned. On the face of it, nothing is more permanent than a mountain range. Perhaps equally enduring in appearance is the Grand Canyon. It was to be precisely these exemplars of permanence which would provide the evidence that our planet had not seen one moment of stability in its entire history.

This Wreck of a World

The prevailing view as to the permanence of the physical universe, the earth included, was occasionally challenged by geologists during the 17th century. The idea that the earth had undergone significant change since its creation was a disturbing thought, and it was vigorously attacked. In 1695, John Woodward, Professor of Medicine at Gresham College in London and an avid collector of fossils, felt constrained to put his thoughts into an *Essay Towards a Natural History of the Earth*. He stated his conviction that the earth we see now is the earth that was originally created:

> ...[T]he Terraqueous Globe is to this Day, nearly in the same Condition that the Universal Deluge left it; being also like to continue so till the Time of its final Ruin and Dissolution, preserved to the same End for which 'twas first formed.

This statement communicates the traditional view and its values. Woodward acknowledged that the Universal Deluge (the Biblical flood) affected the original earth but, since that time, he insisted things have been essentially stable. The statement reveals why the stability of the earth was so important to Woodward. Our planet had been formed by a beneficent God for the purpose of being the place where His creatures would live and flourish until the time He decided to end its use. Any proposal that this world, with its intentional design, had undergone massive changes was impossible to accept.

Two strong forces were in operation, however, which would combine to challenge the permanence of the earth. The first was the application of Newtonian thought to geology. The second was the development of a time scale which expanded the age of the planet beyond the range of normal human thought.

The cataclysmic deluge explanation had been a satisfactory explanation for the appearance of massive geological features such as canyons, deserts, and mountains. Much more subtle and gradual kinds of events were more satisfying as explanations for less massive observable changes. Erosion had been documented throughout history as had the effects of earthquakes and annual flooding of major rivers. To anyone willing to acknowledge the worn appearance of stone exposed to generations of footsteps, it was possible to communicate the potential of gradual and constant effects over extremely long periods of time.

The Uniformity of Natural Events

During the 18th century the combination of the constancy of Newtonian laws acting over very extensive periods of time set the stage for a perception that it was not necessary to appeal to isolated catastrophic events as explanations for even the most monumental geologic features. The deepest gorge need not have been channeled by a single cataclysmic shock; it could have been cut by the wearing action of eons of spring floods. A desert need not have been swept clean of all evidences of life by a gigantic flood but by the slow cumulative effects of climatic change. The mountains, their peaks enriched with fossil sea shells, need not have been wrenched up from the oceans overnight. Their rise out of the depths might have taken unimaginable stretches of time, inch by inch as the crust of the earth slowly and ponderously shifted.

Uniformitarianism became the watchword: the observable natural phenomena of daily experience acted in the past in the same manner they did at the present. Given sufficient time, these natural forces could entirely reshape the globe.

Buffon championed the uniformitarian explanation as an alternative to the traditional **catastrophism** which he saw as totally incompatible with the grandeur of the constantly acting Newtonian forces. The first three volumes of his *Histoire naturelle* were uncompromisingly uniformitarian. His fourth volume, however, reveals the displeasure of the royal censor. In its preface, Buffon retracts what were deemed his heretical views and states

" Pure supposition was an acceptable academic involvement."

his total and sincere belief in the specific wording of scripture. As was typical in his time, he wrote that any alternative ideas he presented were solely on the basis of "pure supposition." Everyone was satisfied. Pure supposition was an acceptable academic involvement.

A Succession of Worlds

Speculation concerning the processes that had shaped the planet was fueled by the findings and speculations of geologists and mining engineers. They recognized three kinds of mountains. The first kind are the lofty peaks such as in the Alps where the very hard rocks consisted of highly compressed and deformed strata with few if any recognizable fossils. It was in these rocks that most of the valuable metal ores were found. These mountains consisted of sharply inclined layers, and if these layers were followed to the base of the mountains, information from deep wells revealed that they plunged underground where the layers became horizontal and extended for unknown distances under the adjacent plains.

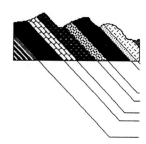

A second class of lesser mountains and foothills on the flanks of the giant peaks also consisted of layered strata. These strata were softer and less highly compressed and were the source of most fossil deposits. These layers also disappeared into the earth where they lay horizontally on top of the deeper, highly compressed strata. In thinking about what they had learned, it appeared to the geologists that sometime in the past the two originally horizontal layers had been uplifted in gigantic folds which had subsequently shattered and eroded. The softer rocks of the upper layer had eroded away exposing the lower layer's harder rocks. What had been the lower layer was now left as peaks sticking up through the eroded gaps in what had been the covering layer.

The third class of mountains appeared to be of volcanic origin since all active volcanos were of this type. The rocks of these mountains lacked regularly layered appearance and gave evidence of having been formed from lava flows and ash from the earth's interior.

In addition to producing mineral wealth, mining exposed the deeper internal structure of the earth's crust. Similarly, whenever a road had to be cut through rock, the patterned interior of the crust was revealed. The roots of mountains were encountered in tunnels, and, in particularly revealing locations, an astounding discovery was made. The badly eroded peaks of Alp-like mountains were found *underneath* other mountains whose peaks were now lifted in snowy grandeur. Layers of volcanic material, evidence of the fiery interior of the planet, were discovered sandwiched *between* layers of stratified rock. Mountains seem to have come and gone again and again leaving their worn-down remnants in layered evidence of an inconceivably long history.

"The erosive capability of wind and water would do the job if sufficient time was available. Could the earth possibly be that old?"

The eruption of a volcano was a known and overwhelming reality. Active volcanos threatened cities rebuilt over the entombed remnants of their predecessors. But to find mountains on top of the remnants of mountains demanded the identity of a force powerful enough to level the former peaks. The erosive capability of wind and water would do the job if sufficient time was available. Could the earth possibly be that old? Could the gradual forces of wind, water, and gravity wear down the Alps and bury the debris which would again be compressed into rock layers which would be uplifted yet again only once more to be worn away? Was reality a cycle of formation and decay? What kind of time scale would be required for such an explanation?

No Vestige of a Beginning, No Prospect of an End

Buffon died in 1788. In that year James Hutton (1726-1797) presented a theory which took Buffon's uniformitarianism to its ultimate expression. Hutton's *Theory of the Earth; or an Investigation of the Laws Observable in the Composition, Dissolution, and Restoration of Land Upon the Globe* was an uncompromising challenge to the belief in a stable and unchanging world. The biblical description of creation, said Hutton, was incapable of accounting for the evidence provided by an examination of the earth.

> For having, in the natural history of this earth, seen a succession of worlds, we may from this conclude that there is a system in nature; in the manner as, from seeing revolutions of the planets, it is concluded, that there is a system by

which they are intended to continue those revolutions. But if the succession of worlds is established in the system of nature, it is vain to look for anything higher in the origin of the earth. The result, therefore, of this physical inquiry is, that we find no vestige of a beginning, - no prospect of an end.

This statement has a chilling quality which emerges from the rational extension of Hutton's premise. If there is a "system" in Nature, in other words, a set of fundamentally applicable laws, that system applies to all of Nature. We see the cycles of the planets, their revolutions, year after year. Taking this as an indication of the importance of cycles and extending that to the apparent succession of worlds revealed by the layers of former mountains, Hutton comes to the conclusion that the world has been cycling from all time and is destined to cycle forever.

"...Hutton comes to the conclusion that the world has been cycling from all time and is destined to cycle forever."

It should come as no surprise that Hutton was attacked by his more conservative scientific contemporaries as having served the "infidel purpose" of discrediting scripture. Among many obvious points of conflict, the age of the earth as deduced from scripture was an impossibly short period of time for the accomplishment of the events of the geological speculations. Biblical scholars, using the "generations" of the Old Testament, had estimated that 6000 years was the approximate age of the earth. Untold millions of years appeared to be needed to permit uniformitarian processes to cycle the succession of worlds revealed by the layered crust of the earth.

Following Hutton's death in 1797, a vigorous defense of his theory was mounted by the mathematician and natural philosopher John Playfair (1748-1819). In an attempt to bridge the gap between the biblical 6000-year figure and the inconceivably longer time required by the geological account, Playfair proposed that the 6000-year figure applied only to the length of time humans had been in existence. Prior to the appearance of Adam and Eve, Playfair suggested, the earth and its living creatures had, as Hutton stated, been cycling innumerable times as world after world came and went.

This approach opened up an even more devastating interpretation of events. In all of these former worlds

"Were there former people and cities and empires in untold shadowy succession stretching back through infinity?"

were there animals and plants which lived and died with their mountains? Were there former people and cities and empires in untold shadowy succession stretching back through infinity? What kind of a god would create such a universe? It was certainly not the loving God of the Christian world.

Playfair compared Hutton with Isaac Newton. In establishing a system for the behavior of matter and motion in the universe, Newton had prepared the way for Hutton's use of the Newtonian forces in discovering a system for the behavior of the earth. Both men had presented their fellow human beings with a vision of reality which was very difficult to reconcile with all of the values of antiquity. As we have seen, these two were far from the only ones whose thoughts were bringing the old and the new into conflict. Greek thought and Christian revelation were not readily meshed with the mechanistic way of questioning.

This does not mean that attempts were not made. The closing years of the 18th and early years of the 19th century produced a variety of proposals which took pieces of information from the older traditions and related them with the new thinking. Buffon, who had been so instrumental in furthering mechanistic interpretation, worked out a system of *epochs* which were distinct periods of time during which various geological events were assumed to have occurred. His first epoch was the time when the earth was fused, by fire, into a molten ball. His second had the ball solidify into rock. The third had water cover the earth, and it was during this epoch that the great limestone deposits were formed by the deposition of innumerable tiny skeletons of marine organisms settling to the floor of the seas. Buffon's fourth epoch had the waters withdrawing to expose the land and his fifth and sixth epochs explained the distribution of living creatures to various locations on the planet. Six epochs and six days of creation were hardly a coincidence. Buffon's epochs were many thousands of years long. His efforts failed to impress the geologists since their concepts required incalculable lengths of time, and his conservative colleagues were outraged at his obvious manipulation of the biblical time scale.

The Fossil Record

These three words carry a series of implications which we will have to consider before we can understand their true meaning. First of all, in order to think about fossils at all requires that one be satisfied as to what they are. By the close of the 18th century many naturalists were in agreement that fossils are indeed the remains of organisms which really lived and died and whose remains have become petrified. The organic materials had been replaced with mineral deposits which faithfully recorded and preserved the structure of the original.

"It is hard not to covet a piece of rock which clearly shows the ridged bark and concentric growth rings of a tree which lived more than 200 million years ago."

As to the word "record," we must ask, *what* has been recorded? We have all seen a display of fossils. Perhaps we have a collection on a shelf at home: a fossil shark tooth, the impression of a leaf, a piece of polished stone with innumerable small shells jumbled together. Some of us have been to the Petrified Forest National Park in eastern Arizona and have walked among the shattered trunks of stone trees. It is hard not to covet a piece of rock which clearly shows the ridged bark and concentric growth rings of a tree which lived more than 200 million years ago.

In one sense, any group of fossils establishes the fact that some creatures existed at some time in the past. But that is not much of a statement. We are not speaking of the fossil collection or the fossil accumulation. In general, when we use the term "record" we have in mind evidence of a sequence of events. The Department of Records at City Hall provides us with the births, deaths, marriages, sale of homes, and all of the other events in the historic life of the community. Because the documents involved are dated, we can reconstruct the sequence of happenings. We can establish family trees by noting marriage licenses and birth certificates. Records, if they are *dated*, permit an accurate recreation of historic sequence.

To see the full implication of this statement, imagine that the Department of Records did not put the date on any of the documents. Would it be possible to accurately establish the sequence of events? From the undated birth certificates, all named Smith, could you tell which individual was the oldest? Could you establish clearly which of the Smiths were the parents and which the children?

Pursuing this metaphor will lead us into interesting revelations.

If the Keeper of Records filed all the certificates *in sequence* even though none of them had "absolute" dates we would have "relative" information. We'd know which piece of paper was the first placed in the file, which was second, and so on. We would never know whether the first birth occurred in 1894 or 1928 but we *would* know that the name of the first baby born was Frank Smith. Sequential filing would give us a way to break into the dateless puzzle. With a lot of ingenuity we would be able to reconstruct quite a bit of reasonably accurate history. Everything would depend upon the records having been scrupulously filed *in sequence*.

Let's imagine that we become insistent about the *absolute* date of the founding of the town. We could examine such things as the handwriting style or the kind of printing used on the certificates. By comparing such features with records kept in towns which did put dates on all documents, we might come up with evidence which would assist us in drawing a conclusion. For example, if a company which supplied printed marriage licenses sold the same product to a number of towns and if this company changed the style of the documents periodically, we would have an *external reference*, that is, a dated document from another town which we could compare with our undated one. Obviously this kind of comparison cannot give total accuracy. Our town may have bought a very large supply of certificates and was still using them for years after the neighboring towns had changed to newer versions. But you get the idea as to how undated information can be interpreted.

How does this apply to fossils as a record? Let's start with sequence. During the construction of Interstate 70 in Colorado, a deep cut was made through the foothills which flank the higher peaks of the Front Range. This kind of terrain was referred to in our discussion of fossil-bearing rocks. The cut revealed the layers of rock so dramatically that it was decided to build a parking area and visitor's center.

I have driven past this spot for 25 years, frequently stopping to experience its power. It never has failed to create in me a feeling of being in a time warp. For when I place my hand against the lowest exposed stratum I am literally touching the stony ghosts of creatures which lived 100 million years before the first dinosaurs shook the earth. My eye travels upward past layers of dark basalt produced by lava flows, limestone deposited on the floor of an ancient sea, green shale indicative of thousands of years of mud deposition, the coal remains of a drowned forest, and sandstone layers of all textures and colors revealing millions of years of erosion of the Rockies whose peaks still shine above me.

Layers of deposited material certify the record of past events as certainly as the meticulously filed but undated documents we used as their metaphor. It was a Danish physician, Nicolaus Steno (1638-1686), who first suggested that stratified rock resulted from the deposition of sediments in a fluid medium. Steno transferred the image of layered mud deposits, which can be observed anytime a turbid stream enters a quiet pond and drops its load of sediments, to the layers of rock such as the ones that can be touched anywhere they are exposed on our planet. Steno had the capacity to see past the scale and time differences to the basic similarity revealed when a shovel slices open the mud layers at the bottom of a farm pond and a road cut reveals the majesty of a cliff of exposed strata in the mountains. The pond's layers were the result of no more than a few years of intermittent and variable deposition of silt, the thin layers resulting from gentle spring rains and the thicker ones from turbulent summer storms. Steno had the mental tenacity to pursue the relationship to its logical conclusion. No matter how high the cliff, its layers revealed that erosion and deposition of unthinkable dimensions in time and space had indeed occurred.

Sediments always settle out in horizontal layers. How does one explain *tilted* strata? The Colorado layers I have described are inclined quite steeply, rising toward the peaks to the west and disappearing under the plains to the east. Steno suggested earthquakes and volcanic upheaval as the forces which disturbed the original horizontal deposits. He used the present appearance to deduce a past history.

This is a critical example of the influence upon scientific

" Science was constrained, by its insistence upon natural law, to abandon the luxury of using mystical inventions to solve its problems."

thought of the change to purely mechanistic interpretations. Steno, having arrived at the deposition explanation, knew that the original position of a layer would have to be horizontal. Gravity acts that way. If the layers are now found to be tilted, he could not argue that they had been deposited on a slant. He had to seek for purely natural forces which could have changed the original horizontal structure. Any buckling, cracking, or slipping out of line of the original deposits had to be accounted for mechanistically. No unknown forces could be invoked. Science was constrained, by its insistence upon natural law, to abandon the luxury of using mystical inventions to solve its problems.

Reading the Rocks

An exposed series of rock layers could be read as a sequential history of former events. The bottom layer in a series had to have been deposited first and thus was the oldest. Successively higher layers revealed later depositions in the historic sequence of events. It was in this context that the fossils took on their critical importance.

Various rock strata contained identifiable populations of fossils which could be studied in the same manner as living creatures; they could be named and classified using the system which Linnaeus had developed. Many fossils were sufficiently similar to living organisms to be accurately placed into existent taxonomic groups. In all cases, however, it was clear that the appearances of presently living creatures had changed from those of their fossil relatives. In order to bridge the gap between the fossil creatures and the living world it would be necessary to accomplish two tasks: the first task was to study fossils with the same level of scholarly intensity as had been expended by the naturalists upon living species. The second was an explanation for the successive changes in the populations of fossils in the layered rocks. The organisms had appeared, remained for a time, and then had vanished never to be seen again. What had occurred to life on our planet?

Georges Cuvier (1769-1832) presents us with the kind of contradiction which makes following the history of scientific thought difficult for modern minds. We anticipate that the development of a subject like science will have occurred as a straightforward series of successive achievements. We are confused by a story which seems to backtrack. Cuvier was a contradictory figure in the effort to make sense out of a most difficult problem.

This nobleman was the leading French scientist of the Napoleonic era. His comprehensive approach to the examination of fossils literally created the science of paleontology — the study of fossils, their identification, their classification, and the interpretation of their relationships. This effort was to result in the recreation of the world which had been and was no longer.

Fossils of all sorts, in remarkable numbers, were unearthed during the latter half of the 18th century. Hunting for and displaying fossils became a mania as the enormous remains of ancient elephants and sloths larger than rhinos were discovered in eroded river banks and reassembled in museums all over the world. Thomas Jefferson became embroiled in a scholarly debate with France's Buffon as to the proper identification of huge teeth found in Virginia. The problem of interpretation was complicated by the fact that skeletons of a number of creatures had been jumbled together by being rolled down both ancient and modern rivers before coming to the location where they had been found.

Cuvier was a comparative anatomist. An isolated bone was an indicator of the entirety of an organism. By using the comparative approach, he was able to move from the known to the possible. If one has a complete house cat skeleton and finds only the lower jaw of a tiger, that isolated bone, by comparison with the jaw of the house cat, enables a vision of a related and much larger catlike animal. By clarifying the patterns of skeletons of present-day animals and comparing them with fossils, Cuvier was able to establish convincing relationships. A gigantic fossil tooth, which was clearly structured along the lines of a modern herbivore, could not have come from the jaw of an ancient lion. Cuvier's scholarly approach established his reputation as the undisputed master of the science he founded.

As he examined the fossils from successively more recent layers, Cuvier saw not only change but *progress*. Certainly the oldest layers contained the fossils of sim-

"So the idea that life had progressed became established in the thinking of those who studied the past. "

pler-looking aquatic creatures. It is only in successively more recent strata that we see the first appearance of fish-like vertebrates, then the first terrestrial animals with complex skeletons, and ultimately, in the more recent strata, the first appearance of mammals. To Cuvier, the progression from simpler to more complex was indisputable. So the idea that life had *progressed* became established in the thinking of those who studied the past.

In my introduction of Cuvier's name to the discussion, I said that he presented us with a contradiction. It comes as a surprise to us to learn that Cuvier opposed the view that the animals in a lower layer were the ancestors of those in an upper layer. In other words, he rejected the possibility that the creatures had evolved.

For Cuvier, the biblical account of creation was as real as the bones in his laboratory. He accepted both as being true, and he attempted to bring his two truths into an acceptable accord. His effort relied upon the acceptance of catastrophism and the rejection of uniformitarian geology. In Cuvier's view, successive rock layers gave evidence of sudden and cataclysmic change which had resulted in the destruction of an existing world, and the dispersion or annihilation of its living population. The next layer provided the evidence of a new creation.

It is his combination of scholarly perception in the paleontological sense and his rejection of the uniformitarian geological explanation which makes Cuvier a puzzle for us. It is valuable for us to learn that science does not occur in a predictably progressive manner. Human minds like ours are what propel science, and we should know full well that these minds are subject to doubt, belief, precision, and confusion. One thing is certain: Cuvier prepared the way for the evidence of the fossils to be applied in a fully mature manner to the overall study of life's history. His conviction as to the progressive nature of that history was to flavor the evolutionary debate which was to come.

The rocks told of indisputable change and demanded a time scale beyond comprehension. The mile-deep chasm of the Grand Canyon revealed, in its walls, worlds upon worlds upon worlds. Mountains are not enduring and seas do not roll forever. The earth was a place of unceas-

"...[We] find no vestige of a beginning, no prospect of an end."

ing change. Ocean depths were uplifted to become mountain peaks, and the peaks were totally eroded to desert flatness, their rock fabric swept away by unimaginable numbers of raindrops down unthinkable numbers of rivers to seas which had not existed but now were there to receive the mountains, grain by grain.

James Hutton said it best: "We find no vestige of a beginning, no prospect of an end."

CHAPTER FIVE

THE CHALLENGE TO CREATION BY DESIGN

*"But however many ways there are of being alive,
it is certain that there are vastly more ways of being dead."*
—Richard Dawkins

I f I ask you this question, "How come there is a tire in the trunk of your car?," your response will very likely be, "It is there in case I have a flat. This is my spare tire for such emergencies."

The Appeal of Teleology

Your perception of my question is that I am inquiring as to the purpose or intended use of the tire. But suppose my question is directed not at the intention of its use but at the mechanism responsible for getting the tire into the trunk. If you understand my question to be of this sort, you will probably say, "Well, someone working on the automobile assembly line put the tire in the trunk."

Before you discard the second interpretation of my question as being superficial, turn things around and imagine yourself inspecting the trunk of a new car which *doesn't* have a tire in the trunk. You ask the car salesman, "How come there is *no* tire in the trunk of this car?" Whereas you might accept an explanation which is of the causal variety, such as "We removed the tire to inflate it properly and apparently forgot to replace it," you would hardly believe that "There is no possibility of your getting a flat so we've discontinued putting spare tires in these car trunks."

Clearly, responses concerning the presence or absence of a tire in the trunk of a car can have been prompted by two quite different perceptions. The first type assumes an *intention* or a purpose for anything as involved looking as a tire, particularly as a component of something else as complicated as an automobile. The aspect of complexity is very important: had you seen a small stone in the trunk you would have viewed its presence as accidental and would not have assumed a purpose. **Teleological** explanations assume that the item or system under investigation has an intended purpose and was designed to accomplish that purpose. The more complex a thing is the more likely it is to have been specifically designed to accomplish a particular intent.

The second type of response assumes that if something is found someplace then there must be a causal *process* which accounts for its being there. Intentions and purposes are not assumed in **mechanistic** explanations which concern themselves only with the processes involved in producing the thing under investigation. This chapter will examine the two different kinds of responses to the question, "How come?"

Natural Theology

It is difficult to discuss teleological explanations without referring to the Reverend William Paley (1743-1805). This English theologian has been quoted so frequently that I hesitate to use him as an example. But Paley was better at stating the teleological argument for creation by design than anyone else, and it would be a disservice to his accomplishment to use a lesser authority.

Paley was continuing an English fascination with the relationships between Nature and God when he wrote his book, *Natural Theology: or, Evidences of the Existence and Attributes of the Deity, Collected from the Appearances of Nature.* The title is self-explanatory. If you want to know if God exists and what He is like, just look at His created beings. They will tell you all about their Creator. In his most famous passage, Paley gets to the heart of the argument for the existence of God.

> In crossing a heath, suppose I pitched my foot against a *stone*, and were asked how the stone came to be there, I might possibly answer, that, for any thing I knew to the contrary, it had lain there forever. But suppose I had found a *watch* upon the ground, and it should be enquired how the watch happened to be in that place; I should hardly think of the answer which I had before given, that for anything I knew, the watch might have always been there.

The complex watch, says Paley, is so different from a stone that its obvious design demands that we come to the entirely different conclusion

> that the watch must have had a maker: that there must have existed, at some time, and at some place or other, an artificer or artificers, who formed it for the purpose which we find it actually to answer; who comprehended its construction, and designed its use.

A stone is simple in that it does not consist of assembled parts. One can accept a stone without asking what the purpose of a stone may be. To be sure, a stone can be used as a doorstop, but nobody has ever seriously proposed that stones were created to be doorstops. Watches, however, are clearly designed to keep time. They are extraordinarily well designed for that particular purpose. It defies belief that they just happen to be good at time keeping. No, they were designed for that purpose, and the springs and hands and numerals on the face are evidence of an intention. The creation, the watch, reveals not only the existence of a watchmaker but also tells us a great deal about the unknown artificer. Paley's watch tells us all we need to know about teleology. The watchmaker creates for an intended purpose, and by examining the way the watch has been designed and made we learn about the maker.

Watches, no matter how skillfully made, pale to insignificance when compared with natural creation. The human eye is a favorite example. Can there be any doubt, says the teleological conviction, that an eye has been specifically designed and made to function for vision? Is the transparent lens not located precisely in the correct position to form an image upon the retina? Can the retina be an accidental thing like a stone? Its various layers of sensitive cells are so precisely connected by neural fibers to specific locations in the brain that it defies reason to assume anything other than painstakingly executed design. The tiny muscles of the iris which control the size of the pupil, allowing only the necessary amount of light to enter, give evidence of an overall appreciation of the role the eye will have to play under varying conditions of light and dark. Nothing has been forgotten down to the protective eyelid and the cleansing tears.

Paley drove home his point:

> There cannot be a design without a designer; contrivance without a contriver... The marks of design are too strong to be got over. Design must have had a designer. That designer must have been a person. That person is God.

Democritus's Fruit of Chance

"Everything existing in the Universe is the fruit of chance and of necessity."

There is a remarkable passage in the writings of Democritus (you will recall that he originated the atomic theory of matter in ancient Greece), where he states, "Everything existing in the Universe is the fruit of chance and of necessity." This statement is, to me, astonishingly wise. I wish it were possible for me to ask Democritus if he truly comprehended the full implication of his assertion.

At first reading, the message is not entirely clear. Let's focus attention on the first part of the statement. "Everything existing in the Universe is the fruit of chance...."

Compare this with the teleological point of view which argues that everything results from an intentional design. Democritus sweeps intention away. For him there is no watchmaker. Everything existing in the universe is the fruit of *chance*.

How can Democritus have come to the conclusion that an eye has come into existence as the fruit of chance occurrences? By "chance" we mean random and unplanned events. Did Democritus really believe that purely random collisions of his atoms would ever produce an eye? The second part of the phrase adds the ingredient "necessity." Everything in the universe results from random events and need. These two don't seem to go together very well. If there is a need it would seem that an item would have to be designed to satisfy that need. The living world has nothing but needs, that is, requirements for structures that see, hear, defend, digest, and procreate. Surely Democritus cannot be serious about random events producing structures which so perfectly accomplish these functions. Let's postpone our judgment of Democritus until we've considered things more fully. But let's not forget his claim.

Darwin's Insight and Its Inspiration

Charles Darwin (1809-1882) is almost always introduced in discussions of evolutionary thought with his trip on the British naval vessel H.M.S. *The Beagle.* He is typically portrayed as a privileged young man with no clear idea of just what he wanted to do with his life. Having dropped out of medical school and studied for the ministry without any evidence of enthusiasm, the opportunity to take a five-year long trip of exploration dropped into his lap. He had an interest in scientific questions but was hardly a trained scientist when he received appointment as the naturalist assigned to the expedition which was primarily a survey of the South American coasts. Most of us know that he spent some time visiting the Galapagos Islands off the coast of Ecuador where he described a group of birds which today bear his name, Darwin's finches. We've been taught that Darwin was a meticulous observer and recorder of the things he saw in South America. It is usually emphasized that he kept extensive notes and collected specimens on the voyage. In addition, it is always stated that the lengthy voyage provided the opportunity to read certain critical books. These factors, we have been informed, set the goals which he pursued diligently for almost a quarter of a century before publishing his book, *On the Origin of Species,* in 1859.

I have an aversion to this portrayal, not because it isn't accurate (because to a considerable extent it is) but because it leads us in the wrong direction in our effort to understand what it is that Darwin said and why what he said has been so central to understanding life. This account emphasizes Darwin the keen observer, the all-encompassing collector of facts, the keeper of voluminous notes. By focusing attention upon the finches and their variously shaped beaks, we may come to believe that it was these birds which made such an indelible impression upon Darwin that they set him to searching for the explanation for their being what they were.

There has been a debate as to which influences came first or were most important. In 1959, the 100th anniversary of the publication of the *Origin*, a number of books appeared which attempted to establish just what went on in Darwin's mind. Since that time, the books have increased in such numbers that someone coined the term the "Darwin industry" to describe the phenomenon. Darwin set the stage for this activity by leaving one of the most massive accumulations of writings ever produced by a single individual. In something akin to biblical interpretation, if you have an opinion about Darwin, someplace in the vast archive of his writings you can undoubtedly find a phrase to support your contention.

The critical thing for us to remember is that Darwin conceived of an *explanatory theory.* The first issue is to determine how one might go about doing such a thing.

in·duc·tion. The act or process of deriving general principles from particular facts or instances.

de·duc·tion. Inference by reasoning from the general to the specific.

Inductive and Deductive Reasoning

There are two basic patterns of thought which are used in almost every scientific endeavor. The **inductive** approach (induction) starts with observations of particular things or events. An example might be a collection of insects; each organism is a discrete thing. The second step in the inductive process is the attempt to detect some unifying generality in the collection. The insects will reveal, to an orderly and disciplined observer, that they all possess an outer skeleton. Chemical analysis of their so-called exoskeletons reveals that all consist of the same kind of complex organic chemical, chitin, which is a nitrogen-containing polysaccharide. From our collection of individual insects we have learned a unifying fact. We might be tempted to see if chitin is present in the exoskeletons of other insects. As we increase our collection, we find that each new individual added does have a chitinous exoskeleton. At some point we feel sufficiently sure of our unifying generality to announce that all insects have such an outer skeleton. We have progressed from a series of particular observations on individual insects to a generalization about all insects.

We can see how this approach might lead us to use other observable features in our collection. We might count the legs of each insect or examine their mouthparts in an effort to extend the list of unifying generalizations. Our intention is clear — to find in the mass of particulars those features which apply generally. By such efforts we hope to reveal the underlying patterns which are sometimes obscured by the bewildering diversity.

"It is a serious mistake to think of Darwin as a collector of facts which, upon examination, revealed to him their pattern of meaning. It was exactly the opposite which occurred. He developed some general principles and then went to the collection of facts to see if they could be accounted for by his concept. This is the process of **hypothetico-deductive** *reasoning."*

The **deductive** process (deduction) also begins with observation. But in this approach the mind comes to the observations of individual things or events with a possible organizing generality already formed. This sounds very "unscientific." It is well that we remind ourselves of the discussion in Chapter One where we encountered Werner Heisenberg's reminder that, "What we observe is not Nature itself, but Nature exposed to our method of questioning." In other words, we cannot truly observe as if our eyes were not connected to our brains. We do have thoughts and attitudes, and it is within this context that we make our observations. Darwin's observations were to be influenced by generalizations, concepts obtained from a variety of sources. Under the guidance of these conceptual generalizations he would interpret the meaning of his observations. It is a serious mistake to think of Darwin as a collector of facts which, upon examination, revealed to him their pattern of meaning. It was exactly the opposite which occurred. He developed some general principles and then went to the collection of facts to see if they could be accounted for by his concept. This is the process of **hypothetico-deductive** reasoning.

Before I discuss Darwin's use of the hypothetico-deductive method in some detail, it is important to acknowledge a confusing aspect of this entire topic. There is an unsettling quality to this induction/deduction business. It is simple to *define* the terms and to give very obvious examples, but there are things which make students wary. First of all, which is the better approach? Have scientists opted for one over the other? In prior courses, students may have been told that science always uses induction because scientists work by assembling individual facts into general theories, never the other way around. As a student I was told that science, unlike philosophy, abhors speculating, yet it is clear that deduction starts with a speculative generalization and then applies that generality to some specific individual case.

I'm going to use an analogy in an attempt to put this into proper perspective. Imagine asking a painter which is better, a broad brush or a narrow brush. The response will be that both brush types are used depending upon what is being painted. Some painters may be inclined to use one or the other more frequently because their style or subject matter dictates a choice. But do you really believe that "painters" would respond that broad brushes are better than narrow brushes?

Scientists use both inductive and deductive approaches depending upon what it is they are working on. They shift back and forth between the two styles of thinking as naturally as an artist makes a brush selection. In fact, just as the effects of each brush stroke will influence the next, so do the results of inductive and deductive thinking influence one another.

Darwin's Use of the
Hypothetico-Deductive Method

Terminology can become either an obstacle to understanding or a convenient labeling system. A great deal depends on the attitude of the reader toward technical language. I don't believe we can avoid using the terminology because the alternative phrase, "if, then," which is sometimes used to explain the hypothetico-deductive method, not only isn't informative but makes the user sound like a kindergartner.

The first part of the label, "hypothetico," refers to the fact that an explanation is proposed as an hypothesis. In Chapter One we encountered Lavoisier's hypothesis that life is essentially a process of combustion, a burning of fuel. Quite a bit was known about the general topic of combustion: it gives off heat, it requires oxygen, it yields carbon dioxide, and it consumes fuel in a predictable manner. So we start with the general category, combustion, with its known qualities and we wish to know if a living thing is a particular case of combustion. Since deduction is a process which starts with a general principle and tests to see if a specific case fits the general principle, we begin to see the reasoning behind this label. Finally, how do we decide if life *is* actually a particular case of a general category, combustion?

We make a prediction. *If* life is indeed a case of combustion, *then* a living thing should require oxygen, yield carbon dioxide, give off heat, and consume fuel at rates similar to other kinds of burning. This is why the phrase "if, then" is sometimes used to describe the reasoning.

Here's an example of Darwin's use of the hypothetico-deductive method. Darwin knew the explanation for the process by which the various breeds of dogs present in his day had come into existence. People had kept records of many breeds, and it was understood that it was possible to identify particular traits, let's say increased body length and short legs, and select dogs with tendencies in these characteristics and breed them to one another. Out of each litter of puppies, the breeder selected those with the characteristics that came closest to the intended outcome and bred such dogs with one another. Wise breeders scoured the country for dogs with the desired traits and did not depend entirely upon their own kennels. This speeded up the process somewhat. Patience and determination, over time, yielded a breed with the desired features. The process was called **artificial selection** since the breeder made all the decisions; the creatures were not permitted a choice. It was artificial in another important sense as well. The kind of animal the breeder wished to produce, the intended outcome of the selection process, was a purely human desire; a long dog with short legs to dig rabbits out of burrows, for example. We know of no wild dogs with such a shape.

Darwin knew that this artificial selection process had produced the various breeds of domestic animals and plants. From each litter or harvest, the breeder selected the particular individuals which displayed the traits of interest and bred these selected individuals with one another. Those animals or plants which lacked the desired traits were not permitted to interbreed with the selected stock. This is an extremely important point. The breeder knew that permitting unselected animals or plants to mate with the selected strains would result in offspring which regressed toward the original traits. For Darwin this was to be a powerful insight. Since isolation of breeding groups from one another was essential to maintain artificially selected qualities, the importance of **isolation** and of **naturally breeding groups** in species formation was to become a major feature of his thinking.

Using the process of **selection** as the **general case**, Darwin speculated as to whether there were any forces operating in **Nature** which might accomplish the same effect as the breeder's decisions had produced in agriculture. His challenge was enormous. The controlled environment of the barnyard or the kennel and the intentional decisions of the breeder were a far cry from the natural world with its blindly mechanistic Newtonian matter and motion. What would play the selective role in the natural world? Darwin was a convinced Newtonian. His hypothesis could not be teleological. There could be no intentions or goals in a purely mechanistic explanation. Yet it was obvious to anyone who examined Nature that animals and plants were ideally suited to the circumstances in which they lived. Furthermore, producing *varieties* of one species, such as the dog, *Canis familiaris*, is one thing; producing entirely new *species* is something else.

So you see, we are embarking on a trip inside Darwin's head. Along the way we may encounter finches and coral reefs and fossil shells on the tops of mountains, all things Darwin saw and thought about. But it is the **theory**, the explanation of the world he encountered, that is our target. Fortunately, Darwin left us a road map with at least a few key names highlighted.

Charles Lyell (1789-1875)

"I never forget that almost everything I have done in science I owe to the study of his great works."

"I never forget that almost everything I have done in science I owe to the study of his great works." Thus Darwin acknowledged his debt to the geologist Sir Charles Lyell. The first volume of Lyell's *Principles of Geology* accompanied Darwin on *The Beagle*. The second volume was shipped to him in South America.

Lyell was a uniformitarian geologist in the tradition of Hutton and Playfair. In reading his book, Darwin developed an understanding of the natural everyday forces which, acting over periods of time beyond comprehension, had shaped the continents which *The Beagle* was circling. There is a series of books entitled *Roadside Geology* which I have referred to while driving through the Rockies. It is remarkable how even a very elementary explanation of the rock formation you are passing adds a penetrating dimension to your gaze. Lyell's book gave added meaning to the sights Darwin encountered.

The book gave him something even more important than geological comprehension: it gave him a model of rigorous scientific thought. Charles Lyell was a disciplined and highly logical mechanist. If he could not develop an explanation based upon purely physical factors, he refused to invoke unknown forces or agencies. In his later efforts, Darwin would be in great need of this example as his theory encountered difficulties which exceeded the scope of his mechanistic explanations.

The Invisible Hand

"One wonders if it was distressing for Darwin to reverse Paley's arguments of a world based on design as a means of pointing out a precisely opposite explanation."

It may come as a surprise that Darwin greatly admired William Paley, the man who had argued so forcefully for teleological purpose. In his autobiography, Darwin praised the logic used by Paley in defending the argument from design. As a young man, Darwin was impressed by Paley's arguments which stressed the precision with which the various "contrivances" (organs and structures) of organisms were suited to the roles that these structures played. One wonders if it was distressing for Darwin to reverse Paley's arguments of a world based on design as a means of pointing out a precisely opposite explanation.

In many presentations of Darwinian thought, Paley is portrayed as a well-intentioned but outclassed opponent. This is a serious mistake. Paley made a number of very perceptive arguments which Darwin used, not as devices to turn against their author, but as positive influences in his own thinking.

In explaining why legs were so admirably suited for walking and eyes so exquisitely formed for seeing, Paley examined the weaknesses in the alternative explanation to the argument from design. If these structures had not been intended for any purpose at all, and had arisen as the result of a random mechanistic process, then only after they had come into existence would the organisms find uses to which their various organs could be put. It would be something like encountering a shop full of tools. You didn't design them, but now that they are available you might discover that hammers are very good at driving nails and saws are excellent for cutting wood.

Paley rejected this possibility. He argued that a leg is fairly crude and might have come about in some random manner. After all, a hammer is not a high-tech implement and can be used to crack nuts as well as drive nails. But the eye? Can one really believe that this incredibly detailed "contrivance," which is so precisely matched to one function and one function only, came into existence accidentally and only then did the organisms discover how nicely this accidental appendage accomplished the job of seeing? This kind of reasoning makes a great deal of sense and it would not be an easy task for Darwin to argue against it.

Paley gave a great deal of thought to the Lamarckian explanation of use and disuse and the inheritance of acquired characteristics. Remember that at the start of the 19th century this idea was an attractive one as a way of explaining the gradual transformation of creatures over time. Since Paley was opposed to the entire concept of change, he took a shot at Lamarck's proposal, and it was a very telling shot.

The Reverend Paley pointed out that in spite of their lack of use ever since Adam, the male nipples were still in evidence and showed no signs of disappearing. Even more devastating to the concept of inheriting an acquired

"...Darwin was tempted to use a Lamarckian mechanism for some details which were otherwise difficult to explain. Paley's ghost would be looking over Darwin's shoulder."

characteristic was the fact that for thousands of years circumcision had been practiced on every male child born into the Jewish faith without any indication whatsoever that the foreskin had been modified in any way. That's what we might call a truly long-term experiment. As we shall see, Darwin was tempted to use a Lamarckian mechanism for some details which were otherwise difficult to explain. Paley's ghost would be looking over Darwin's shoulder.

Typically, when we speak of one person's influence upon another there is an implication that the influence was formative in a positive sense. Paley's influence upon Darwin seems to have been a little different, somewhat like that of an older brother with whom one disagrees but whose opinions are constantly in the back of one's mind.

Paley was a committed defender of the proposition that the world reflected God's designed intention. He explained animal activities as the result of instinctive behaviors with which the Creator had endowed his creatures. In general, he argued, creatures instinctively seek self-gratification. It was relatively easy for him to account for sexual activity in animals; after all, it was pleasurable. But there were some instances of self-denial with which Paley had a problem.

He pointed out that a bird incubating eggs upon a nest is denying all of its self-interest. Birds, he insisted, have instinctive tendencies of flight and activity. What force can be invoked to explain the bird's restraint and "repugnant" immobility?

> For my part, I never see a bird in that situation, but that I recognize an invisible hand, detaining the contented prisoner from her fields and groves for a purpose, as the event proves, the most worthy of the sacrifice, the most important, the most beneficial.

This use of an "invisible hand" by Paley, this appeal to nonmechanistic explanatory forces to accomplish an "important" and "beneficial" outcome was unacceptable to the mechanistically inclined Darwin. But why *do* creatures behave in ways which seem so contradictory to their own self-interest? If there is no invisible hand, then what is there?

Adam Smith (1723-1790)

This Scottish economist played a critical role in Darwin's thinking, and he plays an equally critical role in our efforts to study a science as mature adults. As children, we were taught in what might best be described as the "sheep-herding" tradition. Try to recall a fifth-grade classroom. The teacher had to keep the class from straying off the point because somebody would always attempt to inject an extraneous topic. Children, like sheep, just don't know or care what direction the herder has in mind. Any attractive opportunity seems a good idea to follow. But by the time you finished high school you had been trained to stick to the intended path, and in college you could barely contain your displeasure with the poor fool who was still asking "inappropriate" questions.

"In the middle years of the 19th century, scholars saw the world as much more coherent and less divided into specialized areas of study than we do. Darwin felt entirely comfortable borrowing ideas and applying them wherever they seemed to apply."

It comes as a surprise to find that an economics textbook published in an auspicious year, 1776, but with the improbable title, *An Inquiry into the Nature and Causes of the Wealth of Nations*, could have been a major factor in what we consider to be a purely biological topic.We find it unlikely that economic theorizing and biological concepts have much in common. It is important that we understand a critical difference between Darwin's time and our own. In the middle years of the 19th century, scholars saw the world as much more coherent and less divided into specialized areas of study than we do. Darwin felt entirely comfortable borrowing ideas and applying them wherever they seemed to apply.

What Adam Smith proposed was that the most stable economic system was one in which there was a minimal amount of governmental interference. Laissez-faire (in French, "to let alone") economics argued that if buyers and sellers were permitted to act without hindrance, prices would seek a natural level. If a seller arbitrarily set a price too high, buyers would be unable to afford it, and the demand for the product would fall. Similarly, a product in great demand could be priced as high as the market would bear until the excessively high price inevitably brought down the demand.

Darwin read *Wealth of Nations* while he was still actively searching for the pieces of the puzzle. It is wrong to believe that when *The Beagle* docked Darwin had the questions to ask of Nature all listed in sequence. His writings indicate a much different intellectual process, one which received influences and insights throughout a prolonged period of development.

What Adam Smith did for Darwin was to point out that the behavior of the marketplace was driven by knowable forces. One cannot "see" the law of supply and demand acting to establish prices, as one can indeed see the printed edict of a governmental agency; yet the "invisible hand" of supply and demand functions with equal effect. Darwin believed that the natural world was under the influence of knowable forces, and perceptions like those of Adam Smith became attractive possibilities for his use.

If we can, for a moment, imagine Darwin viewing all the observations and arguments to which his mind has been exposed, arranging the pieces in all sorts of combinations in an effort to have a coherent pattern emerge, perhaps we can appreciate the dimension of the challenge.

The Struggle for Existence

The key to the correct placement of the pieces was to be provided by Thomas Robert Malthus (1766-1834), an English economist, sociologist, and a founding father of the study of population dynamics. In 1798, Malthus published *An Essay on the Principle of Population* in which he observed that populations tend to grow more rapidly than do the means for their subsistence. This was certainly true in the world Malthus knew. Agriculture, the base of the pyramid, was incapable of providing adequate nutrition for a human population which had the potential for exponential growth. A moment's thought reveals his logic.

Each pair of human parents is capable of producing many children. If each of these children, in their turn, mate and produce families of their own, the population potential is enormous. However, arable farmland does not reproduce. As a matter of historic reality, either the family farm had to be cut into smaller and smaller pieces to be distributed among the children, or all but one child had to subsist someplace else. Where else, asked Malthus? On somebody else's farm, he responded, explaining the basis of war. Pointing to periodic plagues and famines, Malthus stressed the harsh reality that populations which had outstripped ther means of production were always vulnerable to devastation by disease and starvation.

There was an additional quality, said Malthus, which had to be borne in mind when discussing the realities of life. Not all people were equally capable of wresting a living from their world. Malthus was not inclined to see a population as some faceless mass of souls all requiring food and drink. He insisted that in *real* populations there were degrees of capability. If food was in short supply, he pointed out, those who were most vulnerable to starvation were the weaker members of the population — the very young and the very old. Living and dying was *not* a game of chance with numbers being drawn out of a hat. People had varying probabilities of staying alive, such as physical strength, prudent and imaginative management of resources, and friends who could assist. Malthus argued that it was these sorts of factors which determined who prospered and who did not. Darwin was to become very conscious of this part of Malthus's argument.

In the writings of Malthus, Darwin saw the outline of Nature's selective force. Adam Smith's balancing of supply and demand were echoed in harsher terms of life or death. We have Darwin's own words for the effect upon his thinking that Malthus' book produced:

> In October 1838, that is, fifteen months after I had begun my systematic enquiry, I happened to read for amusement "Malthus On Population," and being well prepared to appreciate the struggle for existence which everywhere goes on from long-continued observation of animals and plants, it at once struck me that under these circumstances favourable variations would tend to be preserved and unfavorable ones destroyed. The result of this would be the formation of new species.

It was Darwin's "long-continued observation of animals and plants" which made Malthus's arguments particularly meaningful. Darwin was an active research scholar with a deep understanding of the structure and functioning of a variety of organisms. The speculative quality of Malthus's concept found reality in the living world which Darwin had examined for so long.

Here was the critical thread in the tapestry of life. With the perception of a competition for the requisites of life within and among populations of organisms, Darwin had the essential strand which could weave all the others into a coherent design.

CHAPTER SIX

NATURAL SELECTION

"A nutshell definition of science — as of anything else — inevitably floats around on the surface."
—Stephen Toulmin

W hen Darwin discussed a "struggle for survival" among "favourable and unfavourable variations" (we tend to use the terms, fit and unfit varieties) he used language which is inflammatory. The word "struggle" has a violent connotation and the judgmental terms "favourable" and "unfavourable" (or fit and unfit) conjure up racial and ethnic arrogance. This is an unfortunate reality with which we have had to deal ever since his time. But to keep the record straight, we have the obligation to make clear what Darwin was describing.

The Nature of the Struggle

What he had in mind is best portrayed by imagining a backyard in spring. A **population** consisting of a dozen robins is feeding, the robins cocking their heads to look and listen, and then running a few steps to pause and repeat the hunt for worms. Periodically a bird can be observed to stab at the ground; sometimes a worm is captured. The truth is capturing worms is not easy, and there aren't enough catchable worms in this backyard to supply the needs of a dozen robins.

The concept of a **population** is both important and difficult. In biological usage, a population is a group of organisms, all members of the *same* species, which are capable of interbreeding with one another. In our immediate example, it is these twelve robins. They are struggling *with one another* for existence. At issue are worms, appropriate nesting sites, and mates. The worms are limited in this yard; in most years no more than two robins and their offspring can be supported. There are two trees, each with perhaps one truly good location for a nest.

As we observe our 12 birds, we wonder if luck plays a role in finding worms. Does pure chance put a bird in a location where a worm has just come to the surface? Or do certain birds tend to spend more time in locations that favor worm habitation? Are there differences, from one bird to another, in the senses of sight and sound? Or quickness? Unless all of the birds in our population happen to be absolutely identical in every conceivable quality essential for hunting success, Darwin would claim that some were more and some less "favorably" suited in the struggle.

But worm catching isn't the only quality in survival. Choice of nest location and the capacity to construct a nest which will survive summer storms enter into this as well. And ability to attract a mate will determine whether or not a robin will ever breed. Our population has a number of factors operating as it attempts to survive and procreate. As the spring days go by we observe that our population has dispersed to neighboring yards and fields until upon average, one pair of robins occupies and defends against encroachment, its "territory," a piece of the world where it will attempt to execute the business of being robins. Not all the yards are equally suitable for robins. One has a cat. Another has a tree with an exposed and windy location. A third is poor worm country.

One bird may be outstanding at hunting but selects poor nest locations, dooming her construction efforts to repeated failures. Another may be an outstanding nest builder but somehow is just a step too slow in obtaining a meal from the scarce worm population. Another may seem to be acceptable in all ways but for subtle reasons that neither he nor we can fathom, no female will accept him as a mate.

Darwin isn't judgmental; he simply accepts the fact that the birds **vary** in all sorts of ways. On balance, some members of our observed population are better endowed at this particular way of life than others. By the end of summer, all things being equal, the better favored birds will have raised at least one normal, healthy brood. The less favored ones will have raised proportionally fewer survivors. Some pairs with particularly vulnerable nesting sites will have lost all their nestlings to the predation of the ever-present bluejays.

"A Darwinian struggle is not between robins and worms, or robins and cats, or robins and bluejays. It is always within a population, between the members of one species which constitute a breeding group."

The Darwinian struggle for existence occurs *within* a population. It is in this particular sense that species formation can eventually occur. A Darwinian struggle is not between robins and worms, or robins and cats, or robins and bluejays. It is always within a population, between the members of one species which constitute a breeding group.

In order to comprehend the Darwinian explanation for the way the living world is, it is imperative that we understand what Darwin said. **Variation, competition, fitness,** and **selection** are the key words. If we understand their meaning, we can follow Darwin's reasoning.

Variation

Explanations are infinitely elastic; they can stretch as far as you have patience for following them. I am going to state, flatly, that completely understanding evolutionary theory is impossible for someone who has not spent a good part of his or her life in thoughtful consideration of the arguments. Not many of us are that enthusiastic about the subject. But it is possible to get a foot in the door and then to peek inside. That is exactly what we will do.

That members of a biological **population** *vary* is one of the inescapable facts of life. We see it most clearly in our own species, specifically, as we look around the table at a family dinner. Two parents have produced four children. The variation is obvious and at the same time puzzling. Tall, short, dark, fair, male, female, straight-haired, curly-haired, brown-eyed, blue-eyed, on and on go the variations at one dinner table. The same two parents have generated so much diversity.

For most of us, however, the natural world of foxes and crows and maple trees seems remarkably nonvarying. "If you've seen one, you've seen them all" is the typical response to a question about differences in natural populations of animals and plants. It amazes us that out of a flock of thousands of seemingly identical penguins, the mated birds can identify one another and their offspring. Clearly there *is* variation; if the birds were truly identical their identification of mates and offspring could not occur.

The best way to communicate the capacity to identify subtle variations is the experience we may have had with identical twins. I met a pair of monozygotic (from one egg) twins in my sophomore year in high school. I could never be sure which one I was talking to for several months. I realized, a year later, after becoming very close friends, that I no longer thought they looked at all alike. There were significant differences in many qualities; posture and body language, particularly. As these became clear to me, less obvious physical differences became accentuated until I saw them as quite dissimilar. Variation is as real and as pervasive in a school of minnows as it is in a room of eighth-graders.

"Variation is the 'raw material' upon which the process of natural selection operates."

Variation exists in all biological populations. This fact provides the opening insight into the process of natural selection, Darwin's explanation for how species of organisms have evolved. Variation is the "raw material" upon which the process of natural selection operates. When Darwin encountered the finches of the Galapagos with their variety of beak shapes and other adaptive features, he was not discovering variation; he had known of its reality all his life. What was striking was that the variations were distributed among a group of closely spaced islands. The pattern was starkly emphasized and captured Darwin's attention.

Paradoxically, variation, in spite of its key role in the evolutionary process, gave Darwin the greatest amount of trouble. He knew it existed, he understood its importance to his theory, and he invested an enormous amount of effort in trying to arrive at an explanation for it. He, like us, pondered the mechanism that enabled the same two parents to produce such diverse offspring. He died without an answer. We now know this answer.

Competition

The term **competition** communicates this aspect of the process more accurately than "struggle" because it emphasizes that it is between members of the same population that the contest occurs. We are familiar with competition for jobs, for a place on the starting team, and for election to the student board.

I don't want to overemphasize the point that the contestants do not directly eliminate one another because there are clearly some contests in which there is direct violence. But in the overwhelming majority of cases in Nature, it is competition for the essentials of life as contrasted with personal combat which drives the evolutionary process.

Fitness

An organism's fitness is defined as the contribution of its particular genetic content (and the traits that content produces) that it makes to the composition of subsequent generations. Since other organisms are also contributing their genetic content to subsequent generations, fitness is always a relative measurement. Why might one variety of genetic content be contributed in greater abundance than another?

Darwin used the terms "favorable" and "unfavorable" to describe the variations that members of a population possess. It is immediately apparent that such terms require a context: favorable for *what* and fit for which circumstances?

A variation of a trait cannot be simply called "fit" or "unfit." If one member of a population has extremely keen eyesight and another is totally blind, you might at first think that we can forget the context, that is, until I put both individuals in a cave without any light whatsoever. All other things being equal, there is nothing more or less fit about either one. Don't fall into the trap of assuming that the blind individual has some compensating sense which makes it more capable in the dark.

Fitness is entirely a contextual outcome. Any particular variation will be more or less favorable depending upon the conditions under which the creatures live. A long, slender beak may be advantagous for prying insects out from under the bark of trees and is certainly more favorable than a short, blunt one for this purpose. But substitute hard seeds which require cracking and the attributes for fitness are reversed.

We tend to treat the various **adaptations** which creatures possess on an individual basis. In the discussion of beaks or eyes as being more or less fit, there is a danger of losing sight of the fact that it takes an entire functioning organism to *use* a beak. It would hardly be an advantage to have a probing beak which is ideally suited for dislodging insects under tree bark if the bird involved had feet which could not cling to tree trunks. More subtle perhaps, but just as damaging, would be a digestive tract which could not effectively extract the nutrition from insect tissues. Remember that cows can extract all their nutritional needs from grass. We humans cannot survive on such a diet. **An organism's fitness is the totality of all of its adaptations, all of its qualities taken as a whole, within the context of a specified way of life.**

Selection

The birds and the trees can be seen. The various adaptive alternatives of beak shape and foot structure are observable. What is not so obvious is the nature of the process which is at work in determining which of the alternatives are favorable and which are not.

We have been told the Malthusian truth, ever since grade school, that most populations are stable. Our biology teachers pointed out that each pair of fish, for example, produces thousands of fertilized eggs. These eggs produce thousands of tiny fish. But by the end of the year, on average, only two have survived to replace their dying parents. Populations are stable as the result of the balance of two prodigious forces: procreation and death.

Every spring the two maple trees in my backyard release the winged seeds which are the delight of children. Whirling like little helicopters, the seeds spin onto the lawn. The extravagant display goes on for about a week. As I begin to rake them into piles for disposal, I marvel at the productivity of these two trees. Well beyond counting, the seeds have been produced in an abundance which staggers the mind. All over the city maple trees like mine are covering lawns, sidewalks, and streets with untold billions of embryo maple trees, for that is what the bulbous "seed" portion is. Cut open, it reveals the tiny plant inside with its store of nutrition sufficient to enable the root to penetrate the soil and the shoot to rise and unfold the embryo leaves to the sun.

As I drive through streets with windrows of maple seeds and see my neighbors filling trash bags by the hundreds, I am momentarily aware of the two forces. The trees have given, of their substance, tons of embryos. Death lies all around me in the streets of my city.

Some of the seeds take root before I can rake them out of the lawn. One morning, soon, I will look out of the window and see a miniature forest three inches high, a stand of maple trees threatening my grass. The sound of the lawn mower accompanies the decapitation of the entire forest in half an hour. My maples have accomplished not a single replication of their species. Nor has there been more than an occasional survivor, sheltered from death by fortunate location in a hedge, in the city.

Natural Selection as the Mechanism for Speciation

Putting the pieces of the puzzle together was a long and apparently traumatic process for Darwin. As early as 1842 he had written a preliminary version of his theory for the process of species formation. He circulated his paper among friends and associates whom he felt could offer constructive criticism. His subsequent writings trace a path which is marked by delays, detours, and doubts. Why he did not press through to what seems to us to be the obvious conclusion has remained a subject for a great deal of speculation. My own guess is that Darwin knew too much.

If you have a favorite idea and you are unaware of competing concepts, and you are the kind of person who is inclined to move rapidly to conclusions, you will do so. However, if you know a great many things and are aware of many conflicting explanations, it will not be easy for you to dismiss the facts which are difficult to deal with, and it will worry you to turn your back on alternative explanations. Darwin seems to have been compulsively scrupulous about weighing every possibility and giving every doubt its due. At any rate, Darwin did delay for what seems to have been an unusually long time. He was in the process of an interminably long series of revisions and additions to his theory when he was shaken to the core by a remarkable event.

"In what must be one of the most unusual convergences of thought in scientific history, another man had come to the same conclusion concerning natural selection at which Darwin had spent over 20 years of effort."

In what must be one of the most unusual convergences of thought in scientific history, another man had come to the same conclusion concerning natural selection at which Darwin had spent over 20 years of effort.

Alfred Russel Wallace (1823-1913) was a naturalist-explorer who, like Darwin, had spent time in South America. And, like Darwin, he had pondered the remarkable diversity of creatures he encountered. He also had read Malthus.

In a much shorter period of time, several years at the most, Wallace considered the same puzzle pieces and began trying to explain the creation of species. It is reported that in a matter of hours, while recovering from an attack of yellow fever, he wrote a rough draft of his theory. Two days later he finished his paper, a copy of which reached Darwin.

It is impossible to imagine the shock this must have been to the man who had labored so long and so meticulously to assemble his theory. There, in his hand, authored by

Alfred Wallace, was its twin.

The two men published their theory jointly in 1858. Its title was, *On the Tendency of Species to Form Varieties; and on the Perpetuation of Varieties and Species by Natural Means of Selection.*

The Darwin/Wallace Explanation

"Unlike the animal breeder who, using artificial selection, had a conscious goal in mind, natural selection had no intention, no goal to reach."

Darwin and Wallace explained the process as follows. Starting with the unarguable fact of variation among members of a population, they applied the Malthusian assertion that there would be competition among the organisms for the necessities of life. Those variations which enabled effective competition would suvive while less effective ones would fail. The natural environment would accomplish the selection process. Unlike the animal breeder who, using artificial selection, had a conscious goal in mind, natural selection had no intention, no goal to reach. But the effect, in both artificial and natural selection, was the gradual separation, from an original stock, of new varieties and species.

The most critical piece of the explanation involves inheritance. If a variation of a trait, such as a long, slender beak, is advantagous over one which is short and blunt, then those creatures with the "fit" beak shape are more well-suited for life. But in order for their offspring to benefit, the trait must be inheritable. Darwin perceived that it was this aspect of the theory which was critical. If more creatures with the advantageous trait survive and produce young which inherited the trait, eventually the advantageous trait will be possessed by an increasing portion of the population. Successful creatures produce more offspring which, being like their parents, are similarly successful in their turn as parents. Less competitive members of the population, to the contrary, raise fewer surviving offspring, and thus the less successful variation is proportionally reduced in successive generations. It is important to realize that this shift is gradual and occurs over very long periods of time.

A specific example will help us see the major issues. Let's assume that we started out with a population of wolflike animals. The original population became divided into two groups — Population A and Population N.

Population A (for artificial selection) was domesticated by humans starting perhaps 50,000 years ago. The human breeders artificially selected, from among the offspring, those puppies with traits of value to an intended outcome. Large-game hunters desired animals with great size and stamina; others who hunted waterfowl required swimming and retrieving adaptations; and still other breeders selected docility and a smaller stature as being appropriate for companionship.

By preventing the selected animals from breeding outside of their artificially determined populations, over many thousands of years very different creatures were descended from the original wild stock. Irish wolfhounds stand almost five feet tall; Labrador retrievers can swim in frigid waters for hours; and toy poodles become members of a family with less difficulty than the average two-year-old human.

Population N (for Natural Selection) never came under the influence of humans. It was a free-ranging group that spread all over the North American continent. Like all real populations, it had within it variation — in size, in color, and in a thousand other ways. Over the millennia the climate slowly changed in various parts of the continent. To the north the previously mild and wet forested areas gave way to grasslands and ultimately tundra. To the south the almost imperceptible change was leading toward an eventual arid desert environment. Selective forces operated, generation after generation, upon the animals dwelling in the various environments.

The changing climate in the north favored larger size in the prey herbivores: large animals conserve heat better than smaller ones. Also favored are those with a heavier coat, with a minimum of exposed skin through which to lose heat, and with the stamina to cover longer foraging excursions. Selection also favored herd behavior in the large grazing animals. Large groups provided increased protection against the predators.

The colder climate had selected for larger size in the predators as well as their prey. But no individual dog, no matter how large, could overcome the large animals united in a defensive herd. Only predators hunting in packs were successful. Selection favored predators with the innate tendency to function as a group. The northern wolf pack evolved under the selective forces of its environment.

Meanwhile, to the warm and arid south the sparse vegetation and scarcity of water could not support concentrated groups of herbivores. The prey animals had not only become more widely dispersed, but had become smaller; selection had favored the smaller and more solitary variants.

The changes in prey size and distribution had a selective effect upon the predators which also became smaller and more solitary. Quickness, cunning, and the capacity to go for longer periods without food became the adaptations that aided survival. Large size and brute force had become meaningless. And as to hunting in packs, that was a serious disadvantage in the arid south. How many pack members can a single desert mouse feed? No, in the southern desert a solitary hunter, the coyote, a smaller, slimmed-down version with the ability to subsist on insects if necessary, became the favored alternative.

"Darwin and Wallace substituted Newtonian natural forces for teleological intention."

Darwin and Wallace substituted Newtonian natural forces for teleological intention. Natural selection, operating upon the variability that exists in every population, favors the survival of individuals with those traits that are suited to the demands of each environment.

How Do You Explain Variation, Mr. Darwin?

Until we have had the opportunity to discuss genetics (Chapter Ten) we cannot adequately explain several aspects of Darwinian evolution. Variation, which as I have previously said is the raw material upon which selection acts, cannot be understood without a genetic comprehension. Darwin wondered what caused the variations he and everyone else observed, and he flirted with Lamarckian use and disuse for a time before agreeing that he simply had no satisfactory mechanism. It is a shame because the answer was available in his time. Gregor Mendel's insight into the basis of inheritance was published in 1866. Darwin lived until 1882. He cannot be blamed for missing the significance of Mendel's work; nobody else realized its importance until 1900.

How Can Random Changes Result in Improvement?

But there are valid questions which must be asked. We know, today, that variation is the result of purely **random** genetic events. This means that all variations arise due to genetic modifications which follow the laws of chance. Organisms do not have a mechanism for choosing the *direction* of a variation. Even though a longer beak might be a very helpful variation, random events produce all possible variations — longer, shorter, thicker, thinner — without any regard to fitness. Life cannot predetermine the variations it would prefer. It must function with the ones that chance provides.

To many people this seems so contradictory that they cannot believe scientists are so stupid as to believe it. If the variations are totally unpredictable, totally random in Nature, then how do the organisms ever *progress*? How do they ever improve?

Let's examine the improvement of automobiles over the last 100 years. Each modification was an *intentional* change. Instead of a hand crank with which to turn/start the engine, a small electric motor was made to do the work — a "self-starter." The original buggylike structure was intentionally modified for increased comfort, speed, and safety. We understand such modifications as improvements.

Suppose we had *nonintentional* changes — purely random modifications — occurring to the automobile. Each year the factory would pull a change out of a hat and modify the car accordingly. Some possible changes might be more lights, fewer lights, smaller wheels, larger wheels, more wheels, no tires, or fatter tires. It is inconceivable that such a process would produce a progressive improvement.

The problem is our time scale and the numbers of cars we envision being built. We are not referring to a 100-year process but to one of *billions* of years. And not just a few test changes each year are involved. Recall my maple seeds? Each of the billions of maple seeds contained a *random* modification of the parent tree's genetic design. Nature has an alternative to improvement by design. It sounds absolutely mad, but the alternative involves spewing out inconceivable numbers of random changes over time spans which we really cannot comprehend.

The randomly produced modifications are sent out into the world where the ultimate test of suitability is imposed. Death comes to almost everyone. The waste is beyond our ability to discuss rationally. That's because humans operate by *designed* progress. It is not our way to squander treasure blindly. But Nature's approach works. The randomly occurring changes, in numbers beyond imagining and over time beyond belief, are exposed to selection. Those variations which permit survival do so. Those which actually increase survival possibilities become increasingly frequent among the survivor offspring. Those which are unsuited to the world they encounter do not survive.

Biologists accept, as fact, that variations arise randomly. They accept, as fact, that natural selection occurs. They believe that these are the primary factors involved in the process of **evolution**.

What Good Is Half an Eye?

The Darwinian explanation, lacking the guiding hand of design, runs into a resistance which can perhaps best be described as the *withholding* of comprehension. It isn't that we *cannot* follow the argument, just that we choose not to. We simply don't accept that a series of random events can lead to progressively improved adaptations.

One of the difficulties comes from a misunderstanding of how the theory of natural selection explains the evolution of a structure like the eye. Without going into great detail, this misunderstanding looks at the presently existent organ, jumps back to an unknown starting point where there was no eye at all, and imagines a series of changes such that there was, at some time in the process only *parts* of an eye. What would be the survival advantage of an incomplete structure such as a lens but no pupil, or a lid but no cornea, or a tear duct without tears?

A partial eye, according to this view, is of no survival value whatsover, so how does one explain the retention of an incomplete, nonfunctional structure while waiting for the next half-million years of random events to come up with the missing pieces? This is a pretty damaging way to ask the question. It follows from the assumption that eyes are assembled like cars. From this perception it follows that a windshield, without the rest of the car, would hardly be considered a useful adaptation.

But eyes do not form by assembly. A better metaphor would be bread baking. It is highly unlikely that primitive humans refrained from eating grains such as wheat or maize until they had figured out how to bake bread. They obtained whatever nutrition they could from the unground kernels. Grinding the grain into flour increased its digestibility, and mixing the flour with water to form a paste made eating the product easier. The paste, spread on a hot surface, baked into a flat solid. Tortillas, pancakes, and crackers are examples of primitive bread products. They were nutritious, portable, and capable of storage for reasonable periods of time. It is wrong-headed to insist that these primitive forms of baked goods were of no nutritional value whatsoever.

Any organ capable of detecting light is an advantage in a world such as ours. A patch of cells which responds ever so slightly to radiant energy gives sufficient advantage to ensure its selection. A shadow falling on such a patch reveals a change in the environment. It might be nothing more than a passing cloud, but it might also indicate the approach of a predator. If such a patch were to form in a dimple on the skin, it would be a marked improvement since this shape provides an indication of the direction from which the light has come. A deepening of the dimple would be selected for as this would increase the directionality of the patch.

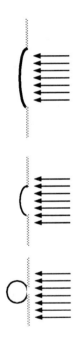

Possible stages in the transformation of a light receptor. Even the most primitive structure confers survival value. Various alternative intermediate conditions would be exposed to selective forces, and successful modifications would be favored in the competition for survival.

We must be constantly careful not to attribute an *intention* to the evolutionary process. The patch of cells has no capacity to project into the future and absolutely cannot direct its changes. It may bulge outward, inward, or not at all with the random shifts of its genetic material. The only thing we can predict with certainty is that it *will* change. And variability, resulting from inherited change, drives the selection process. The realities of the world will determine which variations are suited and will remain, and which are unsuited and will disappear.

How Come All Creatures
Don't Have Eyes Like Ours?

Patches of light-sensitive cells are found on many organisms. Others have simple pits lined with such cells. If eyes like ours are so wonderful, why is it that all organisms didn't develop vision to the same degree that we have?

This type of question reveals as much about our value system as our hesitancy to accept the explanatory ability of natural selection. It is prompted by the same impulse which questions why some cultures continue to bake flat, unleavened bread in wood-fired clay ovens. We take our plastic-wrapped, vitamin-enriched, preservatives-added model as the standard and measure all others against it.

The bread metaphor works quite well, again. To have plastic-wrapped bread requires an entire industrial, economic, cultural environment. Such bread *works* in our culture. It has been selected for against all the other varieties. Its production, marketing, and distribution all *fit* into our society. It takes an entire interlocking culture of trucking firms, banks, advertising agencies, chemical plants, agricultural economists, and Wall Street to permit one loaf to be sold. To demand of all cultures that they have histories, aspirations, and environments exactly like ours is absurd.

So it is with the organisms. A flatworm in the ooze at the bottom of a pond is ideally suited to such a life. It would not find eyes that can outperform those of an eagle a variation of great value. And before you phrase the next question, stop for a moment, and think of your motives. Just because an American living in the final years of the 20th century would rather be an eagle than a flatworm doesn't give that aspiration universal value. No, the flatworm doesn't aspire to fly. It hasn't spent the past million years in the ooze desiring to better its lot. All it has is its environment and a random series of variations. And in the ooze, a flatworm does just fine. Eagles are well-advised to stay out of such places.

Chance and Necessity Revisited

Recall the words of Democritus, written some 2400 years

ago, which are quoted in Chapter Five: "Everything existing in the universe is the fruit of chance and of necessity."

With the Darwinian explanation in hand, we can examine these words from the point of view of random variation and natural selection. In Democritus's world of random collisions of atoms we have the model for the generation of random variations. Blind chance, without intention and purpose, causes alternatives to arise in living things. And what of necessity? Here we find Democritus's vision of selection. What emerges from the blindness of collision, from the chance encounters of matter in motion, is faced with the necessity of existence. What works, works — and survives. What was unable to survive the necessity of reality returns to the chaos of whirling atoms to await another chance collision, another variant incarnation, and yet another exposure to the harsh necessities of this world.

CHAPTER SEVEN

WHAT IS LIFE?

"La fixité du milieu intérieur est la condition de la vie libre."
— Claude Bernard

T he title of the second chapter of this book is "Views of Life." The discussion in that chapter emphasizes the variety of attitudes toward the subject held at different times by different cultures. It is also pointed out that the way in which a question is phrased will dictate the manner in which Nature responds.

It is time to apply the benefits derived from long experience at asking questions, and as François Jacob phrased it, to confront the possible with the actual. In this and the subsequent chapters of this book, we will try to build a representation of life which has the ability to withstand that confrontation. After nearly 5000 years of questioning, what *is* the present state of understanding? What *is* life?

One way of starting our discussion is to point out that until 1800 nobody had ever used the word "biology" to refer to the study of the living state. It was in an obscure medical publication that the term appeared for the first time as a footnote! Two years later, Lamarck used the term and defined it as the study which "includes all which pertains to living bodies and particularly to their organization, their developmental processes (and their) structural complexity." Note that Lamarck *doesn't* include where organisms live and what they eat; in other words, he excludes natural history. The emphasis shifted from the diversity which fascinated earlier scholars to the unifying qualities of the living state.

Aristotle and Linnaeus typify the perceptions of the natural historians whose emphasis had been upon individual living creatures and the variety of ways in which they lived their lives. Darwin, too, began his studies fascinated by life's diversity, and it was only after the lengthy analysis of that diversity that he perceived the unifying forces which shaped the many alternative expressions of life.

But there is a difference between studying the ways in which life *manifests* itself and studying life itself. If we use chemistry as an example I think we can clarify the point.

It is possible to describe water, and sugar, and calcium carbonate, and sulfuric acid, and on and on through all the chemical substances we know. For each one we can list color, taste, smell, and so on. We can describe the uses to which the various chemicals can be put and warn against those which are hazardous. We can provide an enormous body of information. But is that *chemistry*? In other words, have we informed you as to *why* some of these substances will react and others are inert? By listing all the chemical substances are we informed about the forces which hold oxygen to carbon within a sugar molecule? After having all the substances paraded before our dazzled eyes, do we feel that we understand the underlying laws which make some of them corrosive and others mild? By exposing ourselves to the diversity of molecules, have we perceived the unifying principles?

So it is with organisms. They are wonderfully diverse. But after looking at and naming thousands of them, can we explain how they develop from the egg? Can we understand their heredity? Do we know why they age and die?

Now it doesn't follow that simply by using the word biology we gain a totally new and more fundamental understanding. What does occur, however, is by using this word we set our sights on a different goal. Biology is, indeed, the study of the living state, but more importantly, it is the study of those qualities which are shared by all living things and which give to the living state its unique characteristics. Only living organisms develop, have inheritance, are self-replicating, undergo mutation, and evolve by natural selection. Only life operates metabolically, using energy in enzymatically catalyzed pathways of energy flow.

"Beneath the surface of feathers, and scales, and skin, and bark, behind the various ways of behaving, and before the unfolding of leaves each spring and the becoming of embryos each generation there is life. "

Beneath the surface of feathers, and scales, and skin, and bark, behind the various ways of behaving, and before the unfolding of leaves each spring and the becoming of embryos each generation there is life. There is a set of operating rules which apply to all creatures whether they be bacteria or baboons. It is the discovery of these patterns of underlying reality which is the business of biology. It is this realm beneath the surface which we probe when we ask the question, "What is life"?

What Is Life Made Of?

There are really many questions combined in this one. When we ask it with a microscope, Nature responds, "Cells." If we ask it using test tubes, the response is "Various chemicals." So we have to get ourselves under control and be orderly in our questioning or we run the risk of the four blind men encountering the elephant. When one of them touched its tail he announced that an elephant was a rope. A second, encountering the elephant's side thought he was feeling a wall. To the third man the trunk was a snake, and the fourth, wrapping his arms around a leg, believed he held a tree trunk. As with many complex questions, "All of the above" is the "correct" response, but we will learn very little about both elephants and life if we approach the problem this way.

A somewhat different approach is to phrase our question as follows. Is life composed of any substances other than the "ordinary" ones encountered in the **inorganic**, non-living world? The preliminary answer we get from Nature seems to be, yes. Only in living things do we find entire classes of molecules: carbohydrates, proteins, nucleic acids, and lipids. In fact, the name **organic chemistry** was applied to the subject which dealt with the molecules derived from living things. It was believed that life gained its unique qualities from its possession of these "organic" molecules.

It was understood, very early in the 19th century, that the **atoms** (elements) out of which the organic molecules were constituted were quite ordinary. Most organic molecules are composed of the elements carbon, oxygen, nitrogen, hydrogen, and phosphorous. Smaller amounts of sulfur, iron, zinc, and magnesium are involved as are traces of other metals. But there are no elements found in living things which are unique, no special "living" atoms. So why did one not find organic molecules in the inert world? The answer seemed reasonable; only living things possessed the power to assemble carbon, oxygen, hydrogen, nitrogen, and so forth into carbohydrates, proteins, lipids, and nucleic acids. It was assumed that living things had a chemical capability which was not shared by the inert world. This capability, **chemical vitalism**, was seen as a powerful argument in support of the view that life is truly not explainable by using Newtonian mechanism. If only living things possess the ability to assemble the elements into these special "organic" molecules, there must be a **vital force,** in living things that gives them their unique capabilities. **Vitalism**, by definition, is a philosophical doctrine that maintains that life operates by other than the natural laws of physics and chemistry. Those who believe in the doctrine are termed **vitalists**.

"Vitalism, by definition, is a philosophical doctrine that maintains that life operates by other than the natural laws of physics and chemistry."

It is probably impossible to disprove the existence of a vital force. Scientists like to try, however. A favorite effort involves the work of Friederich Wöhler, a German chemist. In 1828, Wöhler synthesized the organic chemical urea. This substance had previously been found only in living organisms. When it was prepared synthetically, in the laboratory, mechanists pointed out that no vital force had been involved.

Scientists are inclined to point to this as a "proof" of the nonexistence of a vital force. A more reasonable statement is that vitalists had been mistaken about organic chemicals. It is true that although they are normally found in living things, they *can* be synthesized in the laboratory as well. But Wöhler's synthesis of urea fell far short of providing proof that living things *lack* a vital force in their preparation of organic chemicals.

What *was* demonstrated was that organic chemicals *can* be produced synthetically. Rather than *disproving* vitalism, it seems to me that what happened was that chemists demonstrated that the vitalistic doctrine was irrelevant to chemistry. They knew that they could synthesize at least some organic molecules and assumed that if they put their minds to it, they could prepare all of them.

The historical process of diminishing a doctrine by ignoring it is what typically has occurred as scientific mechanism matured and earlier explanations were displaced. There have been few serious efforts to disprove the existence of mystical forces; by definition, such forces are not testable under the hypothetico-deductive process. Can you imagine an experiment in which you could convincingly remove the vital force and get no activity and then replace the force and restore what had been lost? How does one put a vital force in a bottle? How does one know that the force is indeed *in* the bottle?

The 19th-century chemists shifted the focus of the questioning by establishing that while organic chemistry was perhaps more complex than inorganic, there were no real barriers between the two. The chemical processes which occurred within living organisms were amenable to study, and the very fabric of life, the stuff out of which it was made, seemed to be chemically identifiable.

Life's Design:
The Concept of Organization

If I tell you that living things are composed largely of water (at least 70% by weight) and that the remaining 30% is predominantly a few inorganic salts with some organic chemicals added, you have gained some information, but in a very real sense you have been misled. It is as if I told you that the ring I intend giving you has a setting which is almost 100% carbon. The setting could be either a chunk of coal or a diamond. Both are made of carbon. The way the carbon atoms are *arranged* or *organized* is the issue.

For two reasons it is misleading to list the chemical constituents of life in percentages. We tend to assign undue importance to larger numbers. Even worse, we assume that somewhere in the list of the substances will be found a clue as to what life really is. We search for meaning in the substances themselves and in their quantities. It is as if life was a cereal box with a list of the ingredients arranged in decreasing order of abundance.

" This concept, organization, is critically important for the understanding of life."

If there is anything about the living state that we know for certain it is that it is the *arrangement* of the constituents that determines the nature of life. **This concept, organization, is critically important for the understanding of life.** It makes no sense at all for someone to tell you that a car consists of 1200 pounds of steel, 600 pounds of aluminum, 300 pounds of plastic, 200 pounds of rubber, and 100 pounds of glass. Without some idea of the way the constituents are organized, this kind of list provides meaningless information. It was not until scientists had an insight as to how to think about the material out of which creatures were made — how to perceive its organization — that a mechanistic explanation of life would be convincing.

Levels of Organization

If I show you a brick wall and ask you to describe it and explain how it was constructed and how its parts relate so as to give the wall its various qualities, you probably will begin by identifying the bricks as the fundamental **units**. The wall is clearly an **assembly**. The concept of **levels of organization** is easily illustrated in such a wall. The individual bricks seem to be at the lowest or first level of organization. If you have ever seen such a wall under construction, you know that the individual bricks are layed in "courses," or layers. So a course of bricks is the second level of organization. All of the courses taken together constitute the third level of organization, the finished wall.

Someone is bound to suggest that the individual bricks are made of *something* and that we had better revise our

Levels of Organization

levels. Bricks are composed of clay which has been fired in kilns. So isn't the clay truly the first level of organization? That same someone will immediately see that clay is also made of something and will suggest that the chemicals which compose the clay are really at the lowest level of organization. Clay is largely composed of aluminum silicates, so have we finally reached our lowest level? No, says our someone, who wants to pursue organization down to the levels of the atoms which constitute the silicates. And then the protons and electrons which compose the atoms. If our someone has had a course in physics, we will hear about quarks out of which the protons and electrons are formed. I think you get the idea.

Now back to life. It seems fairly obvious that the highest level of organization is the individual **organism** — a crow, an oak, a person. Now we begin to see how the concept of **levels** can guide us in considering the way in which a living thing is organized. When we open up a crow we can either consider everything we see as just "stuff," or we can apply some discipline to our examination. There are quite obvious structures — the **organs**. The heart, the lungs, the liver, the brain, the eyes, the stomach, and so forth, are organs. We can assign functions to some of the organs; Harvey helped out with the heart and the blood vessels and the eyes and ears have obvious functions. Others, like the spleen, are not very obvious. But what are the organs made of? What is the next lowest level of organization?

For those of us who have never really examined a stomach, it may come as a surprise that it is not like a brick wall, made of all the same kind of units. It is a multi-layered sac, and each layer is composed of a different kind of material. Using only the most crude magnifying lenses, anatomists had identified various **tissues** and by the close of the 18th century they had established that all organs were formed of combinations of tissues such as muscle, nerve, epithelium, fat, bone, cartilage, blood, and lymph. This realization led to the suggestion that life resides at the level of the various tissues. The so-called **tissue doctrine** was the dominant view among physicians during the early part of the 19th century. They based their diagnoses and treatments upon the concept that all diseases were due to malfunctions in one or more of the types of tissues.

It is perhaps surprising to learn that **cells** had been observed prior to 1665, the year that Robert Hooke brought his observations to the attention of the public in an illustrated book entitled *Micrographia*. The reason I say surprising is that 182 years were to pass before the **cell theory** was formally proposed by Matthias Schleiden and Theodor Schwann in 1847. Along with Darwin's theory of natural selection, the cell theory is one of the two truly significant insights into the nature of life which emerged during the 19th century. We will consider the impact of the cell theory shortly, but we will first examine the circumstances which delayed the realization that it was at the level of the cell that life emerged from the chaotic collisions of Newtonian matter.

Hooke and the other early microscopists were hindered by two quite different obstacles. The first was technological: their microscopes were not up to the task which presented itself. They simply were unable to see sufficient detail with the necessary clarity. The second obstacle relates to Albert Einstein's assertion quoted in Chapter One: "It seems that the human mind has first to construct forms, independently, before we can find them in things." This was undoubtedly the greater obstacle. A cell has to be imagined before it can be seen for what it is.

"A cell has to be imagined before it can be seen for what it is."

When I first began teaching biology I was a laboratory instructor, and many of my students had never seriously used a microscope before. I discovered that it wasn't easy to communicate what I could readily see to people who had no anticipated target for their attention. In other words, their eyes were willing but their brains had not the slightest hint as to how big a cell might be, how sharp its outline would appear, how distinct its nuclear center would be. They looked intently and claimed to see nothing comprehensible. I discovered that it was necessary to prepare their minds by drawing pictures which showed the round field of view under the microscope with the cells drawn precisely to scale and with visual hints as to color, clarity, and detail. Then suddenly, a student would exclaim, "I see it!"

The early microscopists saw fibers, bubbles, webs, granules, boxes, tubes, froth, slime, and so forth. The pictures they drew have meaning for us. We see cell membranes

"For close to 200 years what cells 'really' are eluded the microcopists whose instruments weren't sufficiently developed and whose minds were unprepared to form meaningful images."

and walls, nuclei, cytoplasm, mitochondria, intercellular matrix — all the component parts of a cell population. That's because we have been taught what to look for and what the pieces "really" are. It's the difference between a football game seen through the eyes of an experienced viewer and those of a neophyte. For close to 200 years what cells "really" are eluded the microcopists whose instruments weren't sufficiently developed and whose minds were unprepared to form meaningful images.

The Cell Theory

In the two-year period 1838-1839, Matthias Jacob Schleiden, a botanist, and Theodor Schwann, a zoologist, published a series of reports which defined the central issues which eventually were to be stated as the cell theory. I was led to believe, as a student in an introductory biology class, that the theory emerged as a brilliant stroke of insight and was heralded as a triumph throughout the scientific world. It was disappointing to discover that my teachers hadn't bothered to tell me the "details" of the theory's development. They were eager to have me know the final outcome because that's what we use today to guide our thinking about life. But in not telling me about the birth process of such ideas, my teachers gave me a distorted impression of the way science operates. As a person who would eventually become a scientist, this was not a particularly good way for me to begin.

In plants, living cells secrete quite durable and visible cellulose "walls" which survive the death and decay of the interior material. The early microscopists named the empty spaces "cells." Animal cells do not secrete such walls. Their extremely thin enclosing membranes are neither durable nor readily visible. So there was a lengthy debate among microscopists as to whether "cells" were universally present in living material. Eventually, after the use of preservatives and stains, the microscopists agreed that most organisms they studied seemed to possess such spaces. With improved instruments and techniques, the interior material began to be revealed as well. Most cells had a fairly prominent **nucleus** which typically had a central location in the surrounding **cytoplasm**. The details of the nucleus and the cytoplasm were impossible to discern with any confidence. This led to debates as to roles and origins. Scientists drew pictures, speculated, examined one another's specimens, tried different procedures, and in the end produced an enormous amount of information which lacked coherence. It was this body of work which Schleiden and Schwann, working independently but in constant communication, undertook to examine and organize.

Being excellent microscopists themselves, Schleiden and Schwann were able to check the accuracy of observations, and being well-informed and thoughtful men, they were able to evaluate the speculations. At the outset, however, Schleiden made a very serious mistake which he communicated to Schwann, who fell into the same error.

Schleiden and Schwann held the view that cells came into existence by a stepwise process. In their system, things started out with a structureless fluid or gel which they called the **cytoblastema**. This term means "that which will form cells." The first indication of what was to occur, they said, was the appearance, in the cytoblastema, of a dark central granule, the nucleus. In their minds, the cell came into existence somewhat like a crystal grows in a solution. In a series of successive processes the cell's various structures then appeared until, eventually, the formation of an entire new cell was accomplished.

The reason that this error was so serious is the point of my taking the time to go into the details which my teachers decided not to tell me. In their original version of cell formation, notice that we are left wondering where the cytoblastema comes from. It just is there and begins producing cells. This must be pretty special stuff, this cytoblastema. Had their error not been corrected, scientists might have spent the next 100 years studying something which doesn't exist. There is no cytoblastema.

Fortunately, other investigators had noticed that some cells appeared to have been preserved in the process of dividing when they were prepared for microscopic examination. Eventually, this observation was seen to be the critical step. Cells came into existence by the division of parent cells.

We've come to a very important point in our consideration of life, perhaps the most important one in the

"From the moment the first person wondered about birth and death, life's most mysterious quality is the manner in which the torch of existence is passed from generation to generation. In the cell, science found the unit which was passed. In the dividing of a cell microscopists actually saw the passing of the torch."

entire book. From the moment the first person wondered about birth and death, life's most mysterious quality is the manner in which the torch of existence is passed from generation to generation. In the cell, science found the unit which was passed. In the dividing of a cell microscopists actually saw the passing of the torch.

Life exists at the level of the cell. One cell is truly alive in every sense of the word. It utilizes energy to maintain itself; it contains within its nucleus the genetic information which gives it its identity as a human cell or a cat cell. That same genetic material will be replicated and passed to each of the two cells resulting from cell division, and they will inherit the parental information. No matter how many cells are descended from the one cell, even if an aggregation of billions of cells results in an elephant, the life of that elephant is lived cell by cell. **What cells are and what cells do** *is* **life.**

There is an evasive quality to the last sentence. More exactly, that sentence is a circling definition. Life is defined as what cells are and do, and what cells are and do is defined as life. The statement is true, but it is not satisfying. Exactly what *is* it that cells are and do? Is it possible to escape from the circling definition?

It is possible to describe cells as having membranes and a nucleus and so forth, and it is possible to say that they metabolize and are sensitive to stimuli and to list a host of other attributes. In fact, that is what many teachers insist is the purpose of biology courses — learning the list. But there is another way to break out of the circling definition.

The Milieu Intérieur

A French physiologist, Claude Bernard (1813-1878), was the most influential experimentalist of the 19th century. There is a lot of difference between an observer and an experimentalist. Many of the scientists we have mentioned were marvelous observers: Darwin is a prime example. Many observers are also excellent organizers of their observations: Linnaeus comes immediately to mind. There are also theoreticians, those whose imaginative minds develop possible worlds about which to speculate, and it is their ideas which fuel the experimental process. Claude Bernard was good at all of these approaches to science, but it was as an experimentalist that he had his greatest impact upon his science.

Bernard knew that keen observations were essential to the investigation of life, but he felt that no matter how skillfully one examined organisms it was impossible to arrive at conclusive knowledge by observation alone. Similarly, speculations were sometimes enormously powerful. The atomistic ideas of the ancient Greeks were an example of speculative thought with great explanatory appeal. But without the test of experimental verification — the confrontation of the possible by the actual — the most attractive speculations remained only speculations. Bernard argued that appealing as such thoughts might be, they were of scientific value only if they could lead to experimental testing of their validity.

By using the hypothetico-deductive method, Bernard revealed some of the primary roles of the liver, the pancreas, the sympathetic nervous system, and the oxygen carrying mechanisms of the blood. It was in his consideration of the ways in which the various organs interacted with one another that he developed one of the major insights as to what life really was.

Bernard demonstrated that a variety of substances were produced by the internal organs and that these substances were picked up and carried by the blood throughout the body. Various organs were affected by the substances brought to them. Some substances had a stimulating effect, increasing the activity of the affected organs. Others had exactly the opposite effect, causing a depression in activity. Bernard had a vision of the entire body existing in a state of constancy as the result of the balance struck between stimulation and depression of its many functions. Each substance produced by the body, its numerous hormones, enzymes, even the carbon dioxide resulting from metabolism, was involved in an incredibly complex set of interactions. And what was the outcome of all of these?

We know that our normal body temperature is 98.6° F and that we have a resting heartbeat rate of about 72 beats per minute. Our blood has an acid/base balance which is roughly neutral. Every cell in our bodies is bathed by fluids with precisely the correct salt concentration. What assures this remarkable state of affairs? Even more to the point, when we exercise violently our heart rate goes way up, and the carbon-dioxide level in our tissues rises dramatically, exposing our blood to a potentially devastating acidic condition. The balanced state is on the verge of being tipped out of balance. But the body maintains its equilibrium whether at rest or under the stress of exercise, in fact under all of the varying conditions of day-to-day existence.

"What life was doing every moment of its existence was keeping this internal environment constant in the face of a hostile external world."

Claude Bernard examined the results of his experimental work and the knowledge gained by 1000 years of previous thought and saw in all of this a new vision of life. A living thing, he suggested, is engaged in the accomplishment of establishing and maintaining an **internal environment** (milieu intérieur) which remains **constant** (fixe). All of the activities of the various organs were devoted to the accomplishment of this goal. What life was doing every moment of its existence was keeping this internal environment constant in the face of a hostile external world. If the external environment is cold, life must increase the temperature of its internal domain. If the world was dry, life must conserve its precious internal supply of water. If the critical minerals available from the external environment are too sparse, the inner environment must conserve them. The actions of all of the chemicals produced by the organs of a body and carried throughout the system by the circulating blood are dedicated to controlling the constancy of the internal environment. Life is an island of constancy in a world of hostile change. Each living thing is engaged in maintaining its orderly inner freedom in the face of the chaotic anarchy of the external environment. Between you and me, each of us an ordered and orderly internal world, lies the rest of the universe in its disordered state. As long as life is maintained, as long as our internal constancy of organization exists, we are independent (libre) living things in the surrounding environment of inert matter.

It is from the external environment that life must draw the substances — food, water, and oxygen — of which it is constructed, but life must *organize* these raw materials into a pattern of interactions which enable the maintenance of the internal environment. At death, the *pattern* is lost. The molecules remain but the organization ceases.

Decay is the process of returning the molecules to that external world where they join in the disordered chaotic behavior of the nonliving.

Claude Bernard's vision of physiology, the functions of living things, went well beyond blood pressure or the glucose content of liver cells. These things are some of the measurable manifestations of what a living thing is actually doing. It is certainly true that blood will not circulate unless it is under pressure, and it is one of the roles of the liver to store excess glucose; but Bernard taught us to look beyond the measurable things to their meaning. In times of plenty, glucose is stored. When needed, it will be released to the blood which first brings it from the digestive tract to the liver and which will take it to the various organs as they require it. Blood circulates continuously. It carries instructional molecules which inform organs as to the needs of distant tissues, and it transports substances produced in response to these chemical signals.

The Greeks imagined *pneuma*, a mystical life-giving substance. We have never found any special substance. Instead, we have discovered that life is constituted of very ordinary chemicals arranged in remarkably complex patterns.

The Interior Mold Revisited

It was Buffon who suggested that nourishment acquired by living things had to somehow attain the patterned organization of the living organism. He envisioned an "interior mold" which would impose the proper form upon the raw material. In Buffon's mind, it was the passage of the mold from one generation to the next which was the critical step in explaining life's continuity.

Buffon realized that the word "mold" was only a metaphor and not an actual mechanism. You may recall that when we first considered this subject, I pointed out that the choice of a metaphor was very critical with regard to subsequent investigative efforts. If one is thinking of a mold when observing the various structures and functions of an organism, the eye and the mind are prepared to find moldlike evidences. But, as I suggested earlier, suppose the metaphor chosen was not a mold but a

"The image of a message, of a set of instructions, prepares the eye and the mind to see quite differently than they would in seeking a mold."

blueprint or, better yet, a recipe. A recipe passed from one generation to the next would accomplish the task. And a recipe would not be a structure, like a mold but a message. The image of a message, of a set of instructions, prepares the eye and the mind to see quite differently than they would in seeking a mold.

The final piece which must be put in place in our description of life concerns the possible mechanisms which might accomplish the task of passing the pattern for life from one generation to another.

Informational Molecules

The nature of the intergenerational message turned out to be the most difficult challenge faced by biological scientists. Their first mechanistic efforts were based on the ancient Greek idea of atomic particles. Democritus had suggested that the various kinds of elementary particles would come together to form more complex substances. In the 18th century, Buffon and fellow French scientists Charles Bonnet and Pierre Louis Maupertuis, developed this idea and suggested that the simplest kinds of atoms could, by coming together in various arrangements, form "organic molecules." They envisioned life as consisting of the patterned arrangement of a variety of these organic molecules.

Maupertuis and Buffon saw the reproductive act as the combining of "seminal fluids" from the two parents, and they suggested that these fluids contained all the essential organic molecules necessary to form the offspring. The offspring would, under this model, be formed from a combination of organic molecules derived from the two parents. What was needed, of course, was some kind of instruction as to just how to assemble the molecules.

Maupertuis believed that the each organic molecule possessed a "memory" of its former arrangement and, furthermore, that it had an intention to reform its previous patterning. For him, the assembling of the organic molecules into the required pattern was accomplished by the molecules themselves. Buffon found reliance upon molecular memories and intentions to be too much of a departure from strict Newtonian mechanism. We can see him struggling to avoid the vitalistic implications of

Maupertuis's ideas. His solution was the interior mold, which is certainly a mechanistic-sounding explanation until one begins wondering just how the origin of the mold is to be explained.

It is easy to smile at the efforts of the 18th century until we face the challenge ourselves. Can matter alone, without any externally applied molds or internal but mystical memory and desire, acquire form? And if so, once the form is acquired, can that form be passed down the generations by the matter alone? If there is *only* matter and its motion in the universe, and if all the matter is vibrating and colliding randomly, how can we explain the patterning of life?

If you think about both of these explanations for a bit, you realize that there is a familiar ring to them. They are actually reworkings of the Platonic/Aristotelian duality of matter and form. In ancient Greek thinking, the clay and the potter represented the two necessary but separate parts of the world — the material out of which it was made, and the form which was imposed upon the material to give it its particular nature. For Buffon, the form was imposed upon the matter by an interior mold. For Maupertuis, the physical matter had a nonphysical memory and an intention which guided it to the patterned arrangement.

We have a question to ask of Nature. Can molecules be not only the stuff out of which life is made but also the carriers of the memory of the pattern from generation to generation? Or is life unexplainable under the restrictions of mechanistic thought?

CHAPTER EIGHT

ATOMS, MOLECULES, AND THE ORIGIN OF LIFE

"In reality, nothing but atoms and the void."

— Democritus

"If the atoms have by chance formed so many sorts of figures, why did it never fall out that they made a house or a shoe? Why at the same rate should we not believe that an infinite number of Greek letters strown all over a certain place might possibly fall into the contexture of the Iliad?"

— Montaigne

"Can purely random events, no matter how many occurring for how long, be expected to produce ordered complexity?"

T he two quotations set the stage for the confrontation between the mechanistic and vitalistic explanations of life's origins and fundamental nature. On one hand we have the claim that the world and all that is in it is the product of matter and its motion, nothing else. On the other we are asked if we truly can ever believe that the orderliness of structure and function, the apparent purposiveness of behavior of the living world, can have arisen from a mindless void filled with randomly colliding particles.

1000 Monkeys Typing on 1000 Typewriters

What is the probability that 1000 monkeys typing on 1000 typewriters would produce the complete works of William Shakespeare? Or a million monkeys on a million typewriters? Can purely random events, no matter how many occurring for how long, be expected to produce ordered complexity? The confrontation is ancient and ongoing, but it is not an argument for scholars alone, or for abstract opposing philosophies. Let me pose some questions which concern us as individuals attempting to live our lives in harmony with our views of reality.

When we are advised, by a healer, be that person a western-culture physician or an herbalist practicing oriental medicine, that a particular substance will have a beneficial effect on our health, how do we imagine the substance will act? Whether it is an antibiotic in pill form or an extract of a root in a tea, what do we think is going on as the substance interacts with our bodies? The substance we ingest is clearly matter of some kind. Do we believe that it interacts with the matter of which we are composed, or do we believe that the medicinal matter is not really the active principle at all but only the vessel which contains a spirituous force? Do we envision all healing as the influence of a medicinal vital force upon the vital forces of the living state? Or do we straddle the fence and argue that a medicine, in the form of matter, can influence a vital spirit?

When we express outrage at the release of industrial waste substances which we have been told are carcinogenic and are responsible for leukemia in children, do we believe that the effects are the results of substances reacting with substances? Or do we maintain that the carbon tetrachloride only *represents* the ability to kill? Do we believe that the real power resides in forces which lie beyond the ability of science to detect and know?

These are issues of significance for both individuals and societies. If we believe that the substances themselves are the active agencies then we can approach them directly. But if we think of substances as only the containers of unknowable forces then life forever remains for us an ambiguous mystery.

The Philosophical Atom

I have previously attributed the concept of the atom to Democritus, the 5th-century B.C. Greek philosopher. Actually the idea had been held by his teacher, Leucippus, before him, and we don't really know who had the first glimmerings of the idea. The concept was developed most fully by Lucretius (95-55 B.C.) in *De rerum natura.*

The Greek natural philosophers inferred the existence of atoms from everyday experience. Even though the air cannot be seen, it has the ability to fill a sail. There must be physical reality to it. The water in a heated kettle boils away, and the steam, as it leaves the spout, gives a momentary clue to what is occurring before the liquid water, now vapor, joins the air in invisibility. But the atoms themselves remained mysterious. They were defined as the ultimate and indivisible units. Were they really so? Or do atoms have internal structure which can be subdivided?

"It has been suggested that along with the experimental method and the use of mathematical language, the concept of the atom is absolutely essential for the practice of modern science."

Atomism could be held as a philosophical doctrine, but for most scientists, particularly chemists, the theory was too speculative. They preferred to work with the demonstrable qualities of detectable chunks of matter.

It has been suggested that along with the experimental method and the use of mathematical language, the concept of the atom is absolutely essential for the practice of modern science. It was necessary that a convincing model of the atom be developed, a model which conformed with the ever-increasing number of observations of the way matter behaved under the experimental efforts of physicists and chemists.

The Chemist's Atom

It was the laboratory investigations of late 18th- and early 19th-century chemists, following the insight of John Dalton (1766-1844), which transformed the conceptual atom into a physical one. Atoms were shown to have weight, and further, each element (carbon, oxygen, sodium, etc.) had its own particular weight. When elements combined with one another to form molecules (as when hydrogen and oxygen combined to form water), their individual **atomic weights** were found to be precisely conserved in the **molecular weights** of the compounds. When atoms take on qualities like weight, it is not only easier to think of them as physical things but possible to ask questions about them much like the questions we ask of readily detectable chunks of matter.

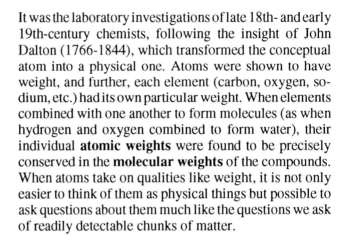

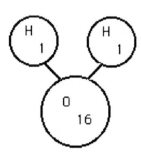

Water molecule consisting of one oxygen and two hydrogen atoms.

For example, hydrogen turned out to be the lightest element and its atomic weight was set at 1. Oxygen is 16 times as heavy as hydrogen and was given the atomic weight of 16. When a water molecule is dissociated into its constituent atoms, only oxygen and hydrogen emerge. A water molecule might logically be expected to have a molecular weight of 17 (1+16=17), but it is actually 18. A moment's thought indicates that perhaps water consists of *two* atoms of hydrogen for each oxygen (1+1+16=18). This idea is testable. Combine hydrogen with oxygen in the ratio of 2:1 and see if this yields the expected amount of water with nothing left over. It does. We represent water as H_2O because of actual results obtained from the testing of speculated combining ratios.

Further insight was provided by the physicists of the 19th century who discovered that atoms could be subdivided and were composed of still smaller component parts, the electrons, protons, and neutrons. In a final realization of the full potential of Dalton's original insight, in 1869 the Russian chemist Dmitri Mendeleev (1834-1907) conceived of the **periodic table** in which all of the elements were arranged according to their atomic weights, the numbers of their protons and neutrons, and the patterns of distribution of their electrons. A periodic table is much like an illustrated list of Lego parts. Some atoms have two connecting locations, some have three, and so on. With all of the elements arranged in groups based on the patterns of their structure it became clear why some atoms were highly reactive while others were almost totally inert. With the periodic table, chemists brought the atom into the realm of orderly thought.

The Biologist's Guide to the Atom: Bonds and Energy

William of Occam (c.1285-1349) was an English philosopher credited with counseling the use of the principle of parsimony expressed as Occam's Razor — "It is vain to do with more what can be done with less." For the intentions of this chapter, we do not need an extensive and detailed account of the atom, just enough to understand a few of the aspects of chemistry that underlie the structures and activities of life.

Understanding the nature of the chemical bond is essential because it is the link that enables individual atoms of carbon, hydrogen, oxygen, nitrogen, and phosphorous to assemble into the molecules of life — DNA, protein, carbohydrate, lipids, and so on. If one is to argue whether or not life appeared on earth by the spontaneous appearance of complex molecules, a minimal understanding of chemical bonds is essential. Otherwise the argument rages on with nobody to referee the debate. There are some rules of the road which atoms and molecules must obey. Not everything is possible.

Electron Rules of the Road

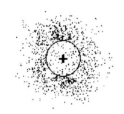

Representation of a central nucleus with a surrounding cloud of electron locations.

There is probably no more familiar 20th-century image than the representation of the atom. Typically there is the centrally located little black ball, the atomic nucleus, and surrounding it the tracks of particles zipping around it in circular paths. We know this is not an accurate representation.

There is a centrally located region in which positively charged protons are located. Also in this nuclear region are the neutrons which carry no electrical charge. The negatively charged electrons do indeed swarm around the atomic nucleus but not in defined orbits. Electrons move in a manner that can best be described as "statistical." We know that 90% of the time any particular electron will be found in an identifiable region called an **orbital**. The rest of the time it can be just about anywhere.

This somewhat casual distribution shouldn't worry you. After all, we live our lives depending upon the probabilities that people and things will be in certain locations at certain times. We have no guarantees of certainty, yet the probability approach permits us to function. So it is with understanding atoms.

The orbitals are actually variously shaped zones extending out from the nucleus at specific angles. It is the shape of these zones and the angles at which they project that establish where the connecting chemical **bonds** can occur. Very much like Lego pieces, not every location can form a connection. The little pegs and their matching holes determine and limit connections. Orbitals define the numbers and locations of chemical bonds.

Energy Levels

The electrons within a given orbital possess a particular amount of energy. Some orbitals are at a higher level, some at a lower level of energy. It is possible for an electron to move from an orbital with a lower energy level to a higher one but only if that electron receives an energy increase sufficient to make the jump. A stone swinging at the end of a string can be pushed to an increased oscillation only if energy is actually transferred to it. Similarly, an electron in a higher energy orbital can move to an orbital with lower energy, but in doing so it must yield the excess energy. If a stone on a string is swinging in a vigorous arc you can make it swing in a much less energetic one by putting your head in the way. The transfer of energy from the stone to your head is very real and demonstrable.

"Life is made possible by the change in orbital locations of electrons. An electron at a higher energy level drops to a lower energy state and gives up the difference in energy to the process of living."

Life is made possible by the change in orbital locations of electrons. An electron at a higher energy level drops to a lower energy state and gives up the difference in energy to the process of living.

Whose Electron Is This?

Ordinarily we think of each atom as possessing a number of negatively charged (-) electrons which balance its positively charged (+) protons. So it is in a *neutral* atom. The number of (-) and (+) charges cancel one another. But it is possible, given sufficient energy input, for an electron to move completely out of *its* orbital and into an orbital of an adjacent atom. Now things are unbalanced. The atom which *donated* the electron is lacking one negative charge and the atom which *received* the electron has one negative charge in excess. The atoms are no longer electrically neutral. The donor has one proton in excess since it is not offset by the missing electron. We call this unbalanced atom a positively charged (+) **ion**. Similarly, the recipient atom has one electron in excess and is now a negatively charged (-) ion. The two ions, being of opposite electrical charge, attract one another. This attraction is termed an **ionic bond**.

This is the situation when a sodium atom (Na) and a chlorine atom (Cl) meet with sufficient energy to transfer an electron from Na to Cl. The Na loses an electron and becomes a sodium ion (Na^+), and the Cl gains an electron thus becoming a chloride ion (Cl^-). The two oppositely charged ions attract one another and the ionically bonded pair forms a **molecule** of sodium chloride (NaCl).

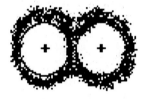

Overlapping orbitals of a covalently bonded molecule.

There are some atomic interactions in which no electrons move from one atom to another; nevertheless the atoms form a bond which is often even stronger than an ionic bond. In such cases, electrons from the two atoms leave their normal orbitals and cooperatively establish a new orbital which is **shared** by the two atoms. Such bonds are called **covalent**, and almost all of the significant biological molecules are formed from such covalent bonds.

There Are Two Kinds of Covalent Bonds: Nonpolar and Polar

If we think of the nucleus of an atom as being positively charged (due to the protons which are located there) and the projecting orbitals as being negatively charged (due to the negatively charged electrons they contain) even an electrically neutral atom becomes a little more complicated than at first glance. While it is true that in such an atom the *number* of protons equals the *number* of electrons, there is the **spatial distribution** of the electrons to keep in mind. In other words, where, in space, are the electrons spending their time?

The simplest atom, hydrogen, consists of one nuclear

proton and one electron which occupies a spherically shaped orbital. No matter from what direction one approaches such an atom, it appears the same in terms of its electrical qualities: a central positively charged region surrounded by a spherical negatively charged one. The atom is not only neutral in charge, but it has no **polarity**; that is, it has no obvious *ends*. It is equivalent in all directions.

When two such hydrogen atoms join by sharing their electrons and thus forming a covalent bond, the two electrons occupy overlapping spherical orbitals and the resulting hydrogen molecule (H_2) is not only electrically neutral (a total of two protons balanced by a total of two electrons) but the molecule is **nonpolar**. The molecular orbital is **symmetrical**; the two electrons are shared equally around the two positively charged nuclei. Such a molecule has no "endedness." In one sense, the molecule is simple. No matter how it is approached its behavior will be the same.

This is not true of all molecules which are formed of covalent bonds. If the shared electrons are *not* evenly attracted to the positively charged nuclei, then their spatial distribution may result in the electrons being distributed **asymmetrically**. Such a case results in the formation of a water molecule.

From its formula, H_2O, we realize that water is formed from two hydrogen atoms and one oxygen atom. Oxygen has eight protons in its nucleus and eight electrons distributed in its orbitals. The eight oxygen protons create a strongly charged positive center compared with the two hydrogen nuclei with only one proton each. When the oxygen atom establishes a covalent (shared electron) bond with each of the hydrogen atoms, the resulting water molecule takes on a distinctive electron spatial distribution. The electrons from the oxygen and two hydrogen atoms redistribute themselves under the influence of the uneven attraction of oxygen's eight protons as contrasted with the single protons in the two hydrogen nuclei. The result is that a water molecule, although *numerically* balanced between protons and neutrons (a total of 10 of each), is a **polar** molecule. The distribution of electrons is not symmetrical. The redistributed electrons form four orbitals. At the ends of two

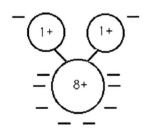

Water is a polar molecule. Its 10 electrons are not symmetrically distributed.

of the orbitals we find the covalently bonded hydrogen atoms. These sites carry a slight positive charge (due to the hydrogen nuclei protons), and at the other two orbitals we find slight negative charges due to the unbalanced presence of electrons.

The Importance of Polar Molecules

As a child I was told that "great oaks from little acorns grow." This proverb was usually meant to convey the idea that if a small amount of money was saved on a regular basis, the result would be a very significant sum. Sometimes this message was applied to ideas, some of which had rather modest beginnings which flourished and grew into dominating influences. But the most striking image of the ultimate significance of small beginnings was captured in this maxim quoted in Benjamin Franklin's 1758 *Poor Richard's Almanac*:

> For want of a nail the shoe was lost,
> For want of the shoe a horse was lost,
> For want of a horse the rider was lost,
> For want of the rider the battle was lost,
> For want of the battle a kingdom was lost,
> And all for the want of a horseshoe nail.
>
> — George Herbert

The polar covalent bond is the horseshoe nail of biological chemistry. The fact that covalently bonded molecules may be polar provides the starting point for what will develop into some rather remarkable molecular behaviors.

The Behavior of Water
as a Polar Covalent Molecule

Nothing seems quite as featureless as a glass of water. Its apparent lack of structure is evidenced by the way water sloshes around and assumes the shape of any container into which it is placed. But when water freezes it assumes a rigidity which reveals that hidden beneath the visible surface is a set of forces which arise from the fact that water molecules are polar in nature. A consideration of these forces explains the transition from water to ice.

As we have discussed previously, the water molecule, with one oxygen atom covalently bonded to two hydrogen atoms, has four orbitals branching off from the oxygen nucleus. Two of the orbitals covalently bond the hydrogen atoms. This end of the molecule carries a slight positive charge. The other two (unbonded) orbitals are negatively charged. When two or more water molecules approach one another the positively charged end of one water molecule attracts the negatively charged end of its neighbor. Similarly charged (+ + or - -) ends of adjacent water molecules repel one another with the result that a population of water molecules takes on a structured quality. Invisible though they may be, water molecules in a glass of water are aligned in a web of attractive/repulsive forces. Water has a crystalline structure. The essence of the crystalline state is ordered regularity. But water is a *fluid* crystal. As long as the temperature of the water is above freezing there is sufficient thermal energy agitating the molecules to keep them from actually locking into the crystal alignment. But once the temperature is reduced below 0°C there is insufficient thermal agitation, and the water molecules do lock into a solid crystal which reveals their polar nature.

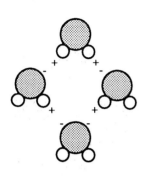

A population of polar water molecules tends toward a structured spatial arrangement.

The polar nature of water molecules influences the way water interacts with substances added to it. A number of familiar organic molecules are also polar in nature; that is, one region of the molecule is positively charged and another is negatively charged. Sugars are polar molecules as are many of the amino acids which form proteins. When polar molecules are added to water, the positively and negatively charged regions of the water molecules align with the charged ends of the added substances. A sugar solution is a complex structured arrangement of the two kinds of molecules. The polar sugar (the **solute**) disrupts the patterned alignments of the water molecules (the **solvent**) by becoming involved in the attraction/repulsion alignments. A sugar solution is far from being featureless.

Similarly, just about any inorganic salt, if dissolved in water, splits into its positively and negatively charged ions. NaCl, for example, dissolved in water splits into

Na^+ and Cl^-. Now consider what this means in terms of the water molecules. Each water molecule has both a positively charged and negatively charged end and the Cl^- ions will be attracted to the positive end of the water molecules. Similarly, the negatively charged end of the water molecules attract the Na^+ ions. In fact, the reason that NaCl dissolves (ionizes) in water is that the attractive forces of the water molcules for the two ions is stronger than the ionic bonds which hold the ions together as NaCl. This is essentially the reason that *any* substance dissolves in water. If a substance is either ionically bonded or is polar that substance, when placed in water, becomes subjected to the polar attractions of the water molecules. A lump of sugar consists of polar molecules. In the dry state all of the sugar molecules are attracted to one another due to their polar natures. They form aggregations called crystals. But when sugar crystals are placed in water it turns out that the polar attraction of water for the sugar molecules is stronger than the attraction of the sugar molecules for one another. As a result, the aggregation is pulled apart, and individual sugar molecules are dispersed among the water molecules. We say the sugar has dissolved.

Were it not for the molecular polarity demonstrated by water and the critically important organic molecules, a good deal of life's structural and functional qualities simply would not exist. It is doubtful that life would exist at all. It is in this sense that the horseshoe nail metaphor is applicable. As we shall see, life's molecular processes hinge upon relationships between molecules, and these relationships require highly specific binding of one substance to another. It is the existence of these critical bonds which identify living processes.

Hydrogen Bonds Function Between and Within Molecules

The easiest way to introduce this particular topic is to return to the polar water molecule, more specifically, to a group of adjacent polar water molecules. Each water molecule consists of one oxygen atom which is covalently bonded to two hydrogen atoms. Recall that the hydrogen "ends" of the molecules carry a partial positive charge while the oxygen "end" is negatively charged. The molecules align themselves so as to bring a positively charged hydrogen close to a neighbor's negatively charged oxygen. The attractive force which exists between the hydrogen of one water molecule and the oxygen of a *neighboring* water molecule is termed a **hydrogen bond**.

It turns out that there are several kinds of atoms which can form essentially the same kind of bond. In all cases a negatively charged region on one molecule attracts a positively charged region of a neighboring molecule. Hydrogen (+) and oxygen (-) are involved in the example that we've used to develop this concept, but hydrogen bonds can form between hydrogen (+) and nitrogen (-) as well.

Hydrogen bonds can form not only between two adjacent molecules but also can link regions *within* one molecule.

It turns out that some organic molecules are so large that regions *within* just one molecule can be polar and therefore possess partial positive and negative charges. The easiest way to portray this is to adopt a metaphor.

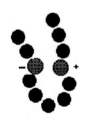

Imagine a string of popbeads. Make the string quite long, let's say 1000 beads. At random, we'll give some beads polar qualities possessing a positive and a negative charge. On average, we've got charged locations every 15 or 20 beads. Now imagine the string of beads free to assume any shape that it can without breaking the string. The string curves gently, and its shape is influenced by the (+) and (-) locations. Opposites will attract and similar charges will repel one another. The string will loop back and forth and will take on a conformation as a result of the attractions and repulsions of its beads. Instead of being straight, the pop-bead string is folded into a shape dictated by the charged locations within it.

There are biologically significant molecules which behave in exactly this manner. Made up of extremely long chains of smaller components, very lengthy molecules, proteins and DNA, for example, take on conformations which result from the locations of charged regions within the chain. Hydrogen bonds are extremely important forces in this process.

Random Behavior Doesn't Mean Everything Is Possible

When we asked the question as to the probability that life could emerge as the result of random collisions of atoms, we used the metaphor of the 1000 monkeys hitting the keys of 1000 typewriters. The metaphor was intended to communicate total randomness in the behavior of the participants, be they atoms or monkeys. There is no guiding hand, no goals or intentions, just matter in motion.

In this era of word processing computers, it is possible that some readers may have never used an old-fashioned typewriter. Basically, such an instrument is a keyboard which is not connected to a memory. On a typewriter, each keyboard stroke is a totally independent event. There is no record, within the typewriter, of what it has done. Any key can be struck in any order. If I happen to hit the "r" key ten times in a row the typewriter faithfully executes my action. The typewriter has no rules it must follow.

But let's consider the actions of a computer. Can you hit any key in any sequence? Indeed you can, but will the computer *execute* faithfully all keystrokes? No it will not. All of us have been beeped at by our computers when we have inadvertently hit a key which had been forbidden by the program. Computers have connections between the keyboard and the memory, the processing unit. There are restrictions imposed on total randomness.

I don't want to extend the typewriter/computer analogy too far before returning to atomic and molecular behavior, but consider this. Can you imagine a software program which restricted the keyboard so that if three consonant keys were struck in a row, the computer would beep and automatically erase the third consonant? It would do the same thing if four vowel keys were struck, erasing the fourth vowel. There could be no sequences such as trm, ntf, aaoi, ioue. It would be possible for a programmer to look at the way English words are constructed and to write a program that would *restrict* certain patterns which do not occur in our language. The person using the computer would be free to hit any keys in any sequence, but the computer would reject any entry forbidden by the program. A monkey could be turned loose on such a machine, and although it could randomly bang away at the keys only certain hits would ever register.

"What we are examining is the idea that random behavior does not mean 'anything goes.'"

Now what has this got to do with the behavior of atoms and molecules? While it is true that atoms are free to collide randomly with one another at all possible angles and at all velocities, the *outcomes* of these collisions are not random. If atoms collide with insufficient energy they cannot react. If they collide at certain angles which do not involve their orbitals, they cannot react. However, if they collide with sufficient energy at the correct angle they may react. What we are examining is the idea that random behavior does not mean "anything goes." There are restrictions which limit the chemical behavior of atoms and molecules. Not everything one can imagine is possible.

Emergent Properties

Very important to our consideration of matter and motion is the idea of history. What does actually happen affects everything from that point on. When the horseshoe nail fell out and made the shoe vulnerable, it set into motion a chain of events that had increasingly significant (and unanticipated) consequences.

A tank of hydrogen gas and a tank of oxygen gas give no indication that if mixed and united they will form a substance which will dissolve salt. Neither one of them alone does so. Neither one of them alone is polar. It is only when the atoms actually have formed water molecules that the idea of wetness could possibly arise. There is nothing about the two gases which predicts that together they could float a rubber duck. There is nothing about the two gases which leads to the expectation that their product will freeze into ice. What we are encountering is the concept of **emergent properties**.

As atoms unite with one another, qualities appear in the products which cannot be anticipated from what one

"It must constantly be emphasized that molecular interactions do not intend *outcomes. Glucose molecules are not trying to become lumber."*

knows about the starting materials. Here's an example. The sugar glucose is quite soluble in water. Individual glucose molecules behave much like table sugar when stirred into water. If one connects two glucose molecules together they form a so-called double sugar (a disaccharide) called maltose. Maltose is also soluble in water. If we continue to add glucose molecules to form a chain (a polysaccharide) we end up with cellulose. This long chain of glucose molecules is not soluble at all. In plants, the cells start out with glucose molecules in solution and connect them into long chains. From what had been a sugary solution emerge fibers of cellulose out of which the wood of a tree trunk is formed. There is nothing about the glucose molecule which would predict that if lots of them were strung together they would make a picnic table.

It must constantly be emphasized that molecular interactions do not *intend* outcomes. Glucose molecules are not trying to become lumber. They aren't even trying to form long chains. But *if* two glucose molecules unite the possibility exists (which had not existed before) that a third glucose might be added. This is what I mean by the statement that history is involved in chemical activity. Things happen, and once they have, even though they are the result of random collision, a new set of circumstances exists. There is no intention and no goal involved, but history cannot be denied. What has occurred will influence subsequent events.

Emergent Properties with Special Significance

We have discussed polar molecules, those which have asymmetrical electron distributions which lead to localized positively and negatively charged regions. Such molecules interact with the polar qualities of water molecules. In general, polar molecules are classified as being **hydrophilic** (having an affinity for water). There are molecules which have entirely symmetric electron distributions and therefore have no localized electrical charge. They are nonpolar. Such molecules have no tendency to interact with water's polar qualities. These molecules are **hydrophobic** (having no affinity for water). This latter group contains the hydrocarbons, molecules which consist of long chains of carbon atoms to which hydrogens are attached. Fats (lipids) and oils are characteristic hydrocarbons. As is well known, oil and water do not mix.

There is a group of lipids which has a peculiar quality. Like most other lipids they consist of two quite long non-polar hydrocarbon chains joined to one another at one end. Attached to this end is a phosphorous-containing component which is polar. This class of molecules, called **phospholipids**, has the remarkable quality of having one end, the two long hydrocarbon chains, which is hydrophobic and the other end, the one with the phosphorous component, which is hydrophilic. This presents the phospholipid molecule with a behavioral dilemma. Should it shun water or seek it out?

We have a molecule with divided affinities. How will it solve its problem? If we pour a population of phospholipid molecules on the surface of water we find that the solution to the problem is very straightforward. The phospholipid molecules all align themselves on the surface with their polar hydrophilic ends down, interacting with the polar water molecules, and their hydrophobic non polar ends sticking up above the water's surface. All parts of the phospholipid molecule, its polar "head" and its nonpolar "tail" behave in accordance with their affinities.

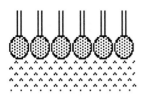

A phospholipid monolayer.

The phospholipid molecules on the water's surface form a **monolayer**, a sheet of molecules one layer thick. In this particular monolayer all of the molecules are oriented in the same direction. Let's stir the water vigorously, breaking up the monolayer into fragments which are forced into the interior. Now the hydrophobic tails seem to have no option but to interact with the polar water molecules. But they cannot; it is not within their chemical nature for the nonpolar tails to relate to the polar water molecules. The polar heads can and do.

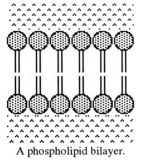

A phospholipid bilayer.

Two alternative structures emerge. The first is a monolayer **vesicle**. This is a sphere in which the hydrophilic heads face outward, in contact with the water, and the hydrophobic tails all are driven into the interior of the vesicle. The monolayer which has been displaced from the surface has rearranged itself into a configuration which satisfies the affinities of its hydrophilic and hydrophobic components.

A second alternative can form. In this one the monolayer becomes a **phospholipid bilayer**. Two monolayers become organized so that their hydrophilic heads face outward contacting the water environment, and their hydrophobic tails are in the interior of the double layer. These phospholipid bilayers also form vesicles but notice that there is a watery *interior* as well as a a watery exterior environment. This is a very interesting outcome. The phospholipid bilayer has trapped a small amount of water inside of the vesicle.

It must be repeatedly emphasized that phospholipid vesicles are the result of purely chemical affinities which arise from polar circumstances in the molecules involved. There is no intention on the part of the molecules to form vesicles. The vesicles are an emergent property arising from the nature of the molecules and their interaction with the environment.

A Leap of Faith?

All cell membranes consist of phospholipid bilayers. The most obvious membrane is the one which surrounds the cell and separates it from the world outside. But there are membranes *within* the cell as well. For example, the nucleus is surrounded with a membrane which separates it from the surrounding **cytoplasm** of the cell. The **nuclear membrane** (or envelope) is also a phospholipid bilayer. Many of the little structures within the **cytoplasm** of cells are similarly surrounded by a phospholipid bilayer. Now it is one thing to demonstrate that bilayers form when phospholipid molecules are stirred into water and quite another thing to speculate that in the origin of cells there was a similar event some time in the very distant past. But that is the way science looks at things. If there are physical laws in operation today, science assumes they have been in operation at all earlier times and places. Imaginative use of today's actualities reveals yesterday's possibilities.

We tend to think of science as a dispassionate and emotionally uninvolved enterprise. Scientists often portray themselves as being detached from their work. If you were to attend a scientific meeting where researchers present the results of their investigations, you would

"Science is about explanation, ideas, hypotheses, and experimental testing. It flows from the creative enthusiasms of its practitioners. Creativity is not a dispassionate quality."

observe behavior which denies this detachment. Science is about explanation, ideas, hypotheses, and experimental testing. It flows from the creative enthusiasms of its practitioners. Creativity is not a dispassionate quality.

Many scientists dislike the word "faith" because of its association with religious conviction, and probably would substitute the word "confidence" when discussing how they feel about their method of study and the information it yields. The reality is that a scientist, like anyone else, adheres to a line of thought because that line of thought is personally meaningful. It has value.

It will probably never be possible to find out if the origin of life did indeed involve the self-assembling of a phospholipid bilayer membrane. What we do know, without a shadow of a doubt, is that such membranes define the borders of today's cells and their internal parts. Also known, without a shadow of a doubt, is that self-assembly of phospholipid molecules into bilayers occurs today if such molecules are stirred into water. Calling it confidence in the thought pattern or faith in the method of organizing information really doesn't matter. What emerges is the scientific approach to explaining the origin of life.

The Origin of Life

"There are only two major alternatives to a purely mechanistic explanation of life on this planet. Either it arose here or it came here frome someplace else."

There are only two major alternatives to a purely mechanistic explanation of life on this planet. Either it arose here or it came here from someplace else. While it is conceivable that some incredibly hardy spore encapsulating a living cell floated here from a distant place in the universe, there is little enthusiasm for such an extraterrestrial origin. I think the basic reason for the lack of interest in this explanation is that it leaves nothing for the creative mind to do. I am aware that all kinds of difficulties present themselves for the successful transport of life across the cosmos — such as radiation, cold, and time spans beyond acceptance — but I think it is the fact that such an explanation leaves nothing for us to do, creatively, that dooms such thoughts.

The Russian biochemist A. I. Oparin published a hypothesis in 1922 which has become the model for a great deal of experimental investigation. It was Oparin's view that life came into existence on this planet by a process of gradual chemical evolution. Starting with simple molecules such as water, methane (CH_4), and ammonia (NH_3), Oparin postulated that more complex organic molecules would form if an energy source was provided. He suggested that the electrical energy from lightning as well as ultra-violet light from the sun and heat from the young earth would be more than adequate to drive the necessary reactions.

In the 1950s, a graduate student at the University of Chicago, Stanley Miller, tested Oparin's ideas. Miller enclosed water, methane, and ammonia in a sealed, heated flask which was bombarded with electrical sparks. At the end of 24 hours, amino acids had formed to the degree that half of the carbon originally in the form of methane was now a constituent of the amino acids. Amino acids are the building blocks of proteins.

In the years since Miller's pioneering work, biochemists have produced not only amino acids but sugars, nucleic acids, lipids, and the like from simple inorganic starting materials. The most elaborate products have been formed in the laboratory of Sidney Fox at the University of Miami. In Fox's preparations, amino acids are the starting materials. Under appropriate heating conditions, the amino acids join with one another to form long chains resembling naturally occurring proteins. If these "thermal proteinoids" are subjected to exposure to salt solutions (an ancient sea?), they form spheres with bilayered membranes.

These experiments, while interesting, provide no evidence that life actually began by such processes. This kind of information provides only possibilities. In the past few years, interest has shifted to chemical evolution occurring at the surface of clay particles, in the vicinity of deep-sea vents, and a variety of other possible locations and circumstances. What all of these approaches share is a conviction that life arose on this planet as a series of chemical steps. The process is envisioned as having taken billions of years; it had no timetable for completion and no road map to direct it. The nature of the atoms involved and the molecules they formed resulted in all sorts of chemical outcomes. Most of the products were degraded back into their component atoms only to be cycled again and again in the chemical chaos of the early earth. At some point, however, perhaps in a microscopic crevice protected from the fury of excessive solar radiation, a membrane may have survived with a watery interior containing a few significant molecules.

What kind of molecules might they be? What is significant for life? What are life's minimal requirements? We have a direction for our questioning, and we have a test for any explanations we may arrive at. If we can account for life's basic attributes by using a mechanistic approach, we will not know for sure that our story is "true" only that it is credible. That's not an insignificant accomplishment.

CHAPTER NINE

THE FIRE FROM A STAR

*"Heat is a motion; expansive, restrained, and acting in its strife
upon the smaller particles of bodies."*

— Francis Bacon

In this quotation from the 16th-century philosopher, we see the effort to apply inductive reasoning to a very evasive concept. By using information gained from particular observations, Bacon attempted to arrive at a generalization. One doesn't have to have been a trained observer to have seen the effects of heat applied to a pot of water closed with a lid. The water gets hot, it begins to boil, the lid starts to jiggle, and eventually it may be blown off by the escaping steam. Heat does indeed show an expansive quality. If the heated substance is restrained there does seem to be action upon the matter in its vicinity.

Bacon attempts to peer beneath the visible particulars and decides that heat *is* motion. In this he gets very close to reality. A pot of cold water and a pot of hot water differ from one another in the amount or degree of motion of the water molecules. In other words, the temperature difference is related to the difference in the molecular motion of the two. If we put a thermometer in the hot water, the fluid in the thermometer expands as its molecules become more agitated, and we measure the expansion as the fluid rises past the marks on the thermometer. Bacon had the intuitive grasp that somehow, in its "strife," heat acted on the "smaller particles" (atoms) and in spite of their efforts to "restrain" it, caused the atoms to increase their motion.

Heat is one of a number of manifestations of **energy**. We are well aware that energy can be exhibited in the form of light, heat, electricity, nuclear reactions, sound, chemical change, and mechanical force. It is a lot easier to measure energy than to comprehend it. For example, if we lift 100 pounds a foot off the ground we have expended 100 foot-pounds of energy to do so. We speak of our cars as having 180 horsepower engines. One horsepower is equal to 550 foot-pounds expended in one second. The point is, we measure energy by the **work** performed. Work is the accomplishment of a displacement (motion from one place to another) of something which resists being moved. It is tempting, therefore, to define energy as the capacity to do work. That which can be measured must be real and whatever has accomplished what you have measured must be real. This is not a particularly satisfying state of affairs to a philosopher who prefers to define things in absolute terms but sufficiently useful for dealing with the world. If you have a 550-pound piano which you want raised one foot and want to do so in one second, you'll need, minimally, a one horsepower device to do so.

Let's assume that the device you have in mind is yourself and three friends. Each of you, stationed at a corner of the piano, will have to exert 1/4 of the 550 foot-pounds during the one-second lift. Each of you must come up with a 1/4 horsepower effort. We know that muscles in your arms, legs, and back will have to provide that energy. Somehow the energy, the force, the "something" which is so hard to envision, must flow in those muscles. Just what is going on as life draws upon this jiggling of atoms, this vibration of matter which powers the muscles of piano movers and the thoughts of philosophers?

The Laws of Thermodynamics

There is nothing more upsetting to nonscientists than a term like "thermodynamics." The word conjures up images of people in lab coats scribbling formulas on chalkboards, formulas which can be converted into nuclear reactions or even worse.

" [T]he laws of thermodynamics explain the reminder that 'there's no such thing as a free lunch.'"

It's a shame that this is so because the laws of thermodynamics explain the reminder that "there's no such thing as a free lunch." The real world in which all of us must earn a living is dominated by the energy laws. When I was a young man I read a newspaper story which claimed that an inventor in Detroit was being persecuted by the oil companies who were frightened to death that his invention would put them out of business. The invention would reconstitute the exhaust gases back into gasoline which would be recirculated to the engine endlessly. The idea seemed so plausible. Why throw away the stuff coming out of the tailpipe? Why not capture it and put it back together as gasoline? The only explanation for the lack of such a device seemed to be the greed of the oil companies which had suppressed the invention. Various versions of the story had the poor guy who had thought the gadget up being held in solitary confinement in some millionaire industrialist's basement or in a federal prison where his economy-wrecking ideas would never be heard. There is something appealing about this story. We want to believe the worst of millionaire industrialists, and we dream of a return to Eden where we will not have to earn our daily bread by the sweat of our brows.

It turns out that this universe runs in a manner which prohibits Eden. I don't know of any universal law which states that greedy millionaire industrialists are inescapable, but I do know that energy behaves in a manner which makes the eternally reusable tank of gasoline impossible.

Energy Conversion

One of the pleasures of a bygone day was the burning of leaves in the fall. After raking the yard I would watch my children jump into the the piles of leaves which would have to be reassembled for the ritual lighting of the match by my oldest daughter. Her younger sisters were kept at a safe distance as the flames spread through the pile, eventually reaching a crescendo of light and heat which forced us all back where we stood in silent contemplation of the death of summer. When the children had tired of standing still and were gone, I would lean on my leaf rake and stare into the flames and think about the hot sunlight of July which had been temporarily trapped in these leaves only to be released once again to heat the chill of October. The leaves had trapped sunlight, and my daughter's match had released the fire of a star.

It is possible to convert one form of energy into another. Electrical energy (which is the movement of electrons) can be converted into light. Solar panels inform us that it is possible to convert light into electrical energy. The energy contained in chemical bonds (as in the leaves) can be converted into light and heat. A fire under a boiler can heat water molecules until they are vibrating so forcefully that they break the hydrogen bonds between them and individually boil out of the fluid as steam. This vibration can push the blades of a huge turbine, causing the tons of metal to spin into a blur of motion which rips electrons from their parent atoms and sends them coursing down a wire as an electric current.

"The First Law also maintains that following the moment of creation, no new energy has ever been created. The universe has a finite and unchanging amount *of energy. "*

The **First Law of Thermodynamics** describes the behavior of energy as it is converted from one form into another. The basic message is that all forms of energy are convertible into all other forms. In addition, when energy is converted from one form to another, all of the energy is conserved. None is ever lost. The First Law also maintains that following the moment of creation, no new energy has ever been created. The universe has a finite and unchanging *amount* of energy.

If this were the end of the story, energy could be thought of like money. Whether in the form of dollar bills, quarters, dimes, or pennies the basic value would be preserved no matter what form of money you had at a given moment. One hundred dollars in quarters is just as valuable as one hundred dollars in dollar bills. But there is more to the energy story.

The **Second Law of Thermodynamics** describes a change in the quality of the energy that occurs during a conversion from one form to another. If we continue the money metaphor, it would be as if you attempted to make a telephone call at a pay phone and all you had were pennies. You have 25 pennies, but the quarter slot pays no attention to your pennies as they roll down the path designed to accept quarters. But in order to understand why some forms of energy are less useful than others in performing work, I'm going to change the metaphor from money to something else.

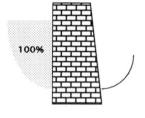

Imagine that you have a million gallons of water behind a dam, all piled up in a reservoir. Below the dam is an empty lake bed into which the water can be discharged. There is a spillway through which the water can pass, and associated with this spillway is a water wheel connected to an electric generator.

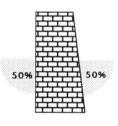

At the start of this process, all of the million gallons represent a **potential** energy source. No energy has actually flowed, and we haven't generated any electricity. When we open the floodgates and the water begins to move, the potential becomes actual in the form of energy of motion, **kinetic** energy. It actually does work, and we start to generate electricity. But the flow of water stops when all the water behind the dam has emptied into the lower lake. The system has reached an **equilibrium**. It is at rest.

Energy distribution before (top) and after (bottom) flow. Following the flow, at the equilibrium condition, the system is no longer capable of performing useful work. The numbers shown are not actual values. Entropy increase has not been represented.

We know, intuitively, that the water will not, of its own accord, ever escape the equilibrium condition. It will not push *itself* uphill. We haven't lost any of our 1 million gallons; it's still all present but it isn't distributed in a manner capable of performing any useful work. There is a sobering lesson here. We have no way of getting any further energy yield out of the million gallons which remain.

The message of the Second Law is that when energy is converted from one form to another, even though there is no loss in the total *amount* of energy, the usefulness of the energy, its capacity to accomplish work, will have been diminished.

Let's revisit the endless tank of gasoline. Gasoline is a hydrocarbon. Its potential energy resides in the chemical bonds which hold its carbon atoms to one another and to its hydrogen atoms. When the gasoline burns, those bonds break with explosive force. Some of the energy emerges in the form of heat. Automobile engines are *hot*! That heat just radiates off as we drive down the road. We are leaving behind a stream of energy that was once in chemical bonds. Some of the energy is kinetic, the movement of molecules, in this case the explosive expansion inside the cylinders. This is what pushes the pistons which are connected to the crankshaft which eventually drives the wheels. The friction of the moving parts inside the engine produces heat which also radiates off. The friction of the rolling tires on the road produces heat. Have you ever touched a tire after a few hours of interstate driving? Every bit of heat that radiates off represents a portion of the original potential energy in the tank of gas. As the car pushed its way through the air it caused friction, mile after mile of heated air molecules to be added into the equation.

The most interesting part comes when we examine the exhaust we've been spewing out for hundreds of miles. The bonds of the original hydrocarbon molecules are now broken. Instead of long chains of energy-rich gasoline molecules, we have little fragments; CO_2, H_2O, CO, and other odds and ends of combustion. If we could capture the heat from the explosion, from the friction, and add it all up and total the shreds of energy still remaining in the H_2O, CO_2, and so on, we'd find that the First Law had been satisfied: all of the original energy is still in existence. But the Second Law has also been satisfied. The energy has been degraded into a form which can no longer accomplish work. The dream of the perpetual motion machine inventor, the perpetual tank of gasoline salesman, is just that, a dream — a fiction.

Entropy: A Puzzling Concept

There is, for most of us when we first encounter this topic, a real difficulty in bringing the two laws of thermodynamics into harmony with one another. The First Law states that energy can be neither created nor destroyed and that the amount of energy in the universe remains forever unchanged. Yet the Second Law insists that when a conversion of energy from one form to another occurs, the capacity to do work has been decreased. If energy is the capacity to do work, how can the ability to do work be decreased if the amount of energy has remained unchanged?

Go back to the dam with the water at its final state. All of the water is below the dam. It is certainly possible to *push* the water back up behind the dam so as to have its potential restored. But it has to be *pushed* back up. That *requires* energy. It will take exactly as much energy to push it back up as was released when it flowed down.

The original million gallons was distributed so that all of it was above the dam. After the water flowed to the lower lake, we still have our original million gallons, but they are incapable of performing any further work.

Physicists have a term they apply this situation. They use the word **entropy** to describe what has occurred. The entropy of an energy system is a measure of its "useless-ness," its *inability* to perform work. In the tank of gas, all of the molecules were capable of burning and accom-plishing work. But when the tank was empty, even though the energy content of the exhaust gases plus all the heat which had been generated added up to exactly the same amount of energy as we started with (the First Law), the form in which the energy existed was incapable of doing work(the Second Law). Similarly, with the dam, the million gallons exists after having flowed downhill, but it is no longer capable of performing any work. In both cases the entropy of the system has *increased*. The *inability* to perform work has increased.

Our Not Quite Garden of Eden

If we think about the lake of water behind the dam, we wonder how it got there in the first place. Melting snow and rainwater were accumulated behind the dam, having run off the sides of mountain slopes above the lake. Trace the rain and snow back one step. They fell from clouds, water vapor condensing into drops or freezing into crystalline flakes. Taking one step back, the clouds formed from the evaporation of water from other lakes and streams and oceans. Molecule by molecule, water was lifted from sea level into the sky. This is a lot of work. Energy was required to accomplish it. Each hydrogen bond which was broken to release a water molecule from its neighbors required an input of energy. There is no free lunch. Where did the energy come from?

The sun shone on the sea and warmed it. The water molecules broke free of the surface. The same heat from the sun warmed the air and made it rise, carrying the water molecules high in the atmosphere. The sun is the energy provider not only for the water cycle but for the life cycle.

My pile of leaves burning so fiercely released sunlight energy which had been trapped in the chemical bonds of organic molecules. When the leaves were green and attached to the tree, water was brought to them from the roots and carbon dioxide from the air diffused into them. The sun's radiant energy was absorbed by electrons which were driven from low-energy-level orbitals to higher energy levels. Chlorophyll molecules in the leaves vibrated at higher and higher levels of energy derived from absorbed sunlight. Their electrons became so energy-rich they broke their bonds and jumped to neighboring molecules until, in a peaking of energy content, they could hold no more and passed the surplus to surrounding H_2O molecules which were split by the force they received. Hydrogen atoms were split from oxygen. The hydrogen atoms were driven into chemical bonds with carbon dioxide molecules to form sugars. Enough energy remained to form bonds linking the sugars together into cellulose. The leaves contributed to the woody growth of the trees which bore them. Our star, the sun, provides the energy which drives the **biosphere**, that thin layer of space at the surface of our planet in which life exists.

*"Our star, the sun, provides the energy which drives the **biosphere**, that thin layer of space at the surface of our planet in which life exists."*

Time's Arrow

We've defined energy as the capacity to perform work. The Second Law states that when energy is transformed from one form to another the entropy invariably must increase. In other words, every conversion of energy reduces the amount of energy in a *form* which can accomplish work. Even though the total amount of energy remains the same, the *useful* portion has been reduced.

Since energy conversions are occurring all the time throughout the universe, it follows that today there is less *useful* energy in the universe that there was yesterday. Tomorrow there will be even less than today. Sir Arthur Eddington (1882-1944), the British astronomer, in referring to this decrease in useful energy in the universe, said "Entropy is time's arrow." If we ever wondered if time does indeed flow, suggested Eddington, all we would have to do is check the amount of useful, work-capable energy in the universe. Entropy's constant increase (and the concurrent constant *decrease* in useful energy) would point the way.

Entropy and Life: Order and Disorder

If one sees leaves in a neat pile in the center of a lawn, the last explanation in the world that would be believed is that a fortuitous wind came through and gathered the leaves from all over the lawn and deposited them in a single pile. It is much more probable that energy was expended in an organized manner to gather the leaves into a pile. However, once a pile of leaves has been formed, it is very likely that a wind will scatter the leaves randomly around the yard. That's simply the nature of things.

We expect random distribution to be the most probable state of affairs. This goes for the position of fallen leaves, rocks which have tumbled down the face of an eroding cliff, the scattering of stars in the sky, the height of children in a second-grade class, and logs which have been rolled off a truck at the site of a cabin under construction. Who would expect to encounter a cabin assembled spontaneously from a jumble of logs? Rocks eroding off the face of a cliff for thousands of years are not expected to neatly spell "Hollywood" as they come to rest. We are particularly good at recognizing **ordered** arrangements as contrasted with randomly distributed or **disordered** ones.

Random disorder is the more *probable* state of affairs. All imposition of order requires an energy input. Rocks which were originally scattered in a field must be put into position to build a wall. If no energy is expended upon them, the rocks simply remain scattered. It is the fate of all structured and ordered states of affairs to fall into randomness unless the ordered state is maintained by an expenditure of energy. This is true for molecules as well as pyramids.

"It is the fate of all structured and ordered states of affairs to fall into randomness unless the ordered state is maintained by an expenditure of energy. This is true for molecules as well as pyramids."

A sugar molecule is an ordered assemblage of carbon, oxygen, and hydrogen atoms. These atoms, originally in the form of randomly colliding water and carbon dioxide, were assembled into a more ordered arrangement. It took sunlight energy, through the process of photosynthesis, to accomplish this conversion. If we leave a sugar molecule alone it will decay back into carbon dioxide and water. An ordered molecule contains, in its bonds, the energy which was used to form it. Decay to a less ordered state releases that energy.

In a very real sense, a living organism is a highly improbable thing. The molecules of which it is composed are each highly ordered, and all of the molecules are assembled into cells — remarkably ordered structures — which are, in turn, assembled into tissues and organs. Indeed, living organisms are extremely improbable. There obviously has been a great deal of energy expended to create the ordered state we see in a living thing.

"There seems to be a contradiction in the highly ordered state of life and the Second Law of Thermodynamics."

There seems to be a contradiction in the highly ordered state of life and the Second Law of Thermodynamics. On one hand we are told that with each energy conversion in the universe, the entropy level increases. Less energy is available to do work today than was available yesterday. The universe is running down. Yet living things are clearly highly ordered states of affairs. Their proteins, nucleic acids, cell membranes, and nerve cells are all much more ordered than the randomly colliding atoms out of which these structures were formed. Is there a contradiction?

Exergonic and Endergonic Reactions

I once attended a lecture presented by a man who told the audience that he had spent many years analyzing the arguments for and against Darwinian natural selection. The speaker's major conclusion was that if physicists were correct about the Second Law then biologists could not be correct about evolution. If the flow of energy was such that the universe was *increasing* its entropy, its randomness, then how was it possible for life to be increasing its orderliness, *decreasing* its entropy? The speaker insisted that science could not have it both ways.

I was seated next to a student of mine who was so outraged at what he perceived to be the speaker's purposeful efforts to mislead his audience that he was getting to his feet to challenge the distortion. I quietly suggested to him that perhaps the speaker really believed that there was a contradiction. The student rejected that possibility stating that a two-year-old could understand what was involved. I urged the student to meet with the speaker privately after the lecture rather than get into a shouting match. The student accepted my suggestion for restraint but wouldn't meet with the speaker. In the student's mind, the answer was so simple he could not believe the speaker didn't know it and was purposefully hiding it from the audience. I'll let you decide.

The student knew that the Second Law applied to the universe as a whole. It is indeed increasing its entropy with every passing second. Each star, including our sun, is blazing away useful energy which, as it is converted inexorably into heat, just dissipates in the random motion of atoms and molecules. There is no way to reverse the overall slide of the universe into increasing entropy and disorder. The physicists are absolutely correct.

But as our sun blazes down, the energy streaming from it can be captured. For example, as sunlight heats the oceans and water molecules break their hydrogen bonds, individual water molecules rise in the thermally heated air. All of this motion is the result of the energy from the sun. Work has been performed. Water is heavy. Try lugging a five- gallon container of it a few miles. When we see a thunderhead towering 30,000 feet in the sky we have evidence of the conversion of light into the energy of position — tons of water raised miles in the air from sea level. When the storm breaks and the rain falls in the mountains we see the rushing torrents answering the pull of gravity as the water pours into a reservoir behind a dam. When released, this water reveals its energy storage by turning the wheel which generates electricity which powers the radio which assures us that the storm will soon pass and the energy-yielding sun will shine once more.

"There is no contradiction between the Second Law's insistence in the universe's entropy increase and the obvious fact that life is a highly ordered state of existence."

There is no contradiction between the Second Law's insistence in the universe's entropy increase and the obvious fact that life is a highly ordered state of existence. The same sunlight energy which raised tons of water miles in the air powers the process of **photosynthesis** in the leaves of green plants. Simple molecules, carbon dioxide and water, are used to form complex, more highly ordered sugars with stores of energy trapped in chemical bonds. The energy-yielding sunlight has been utilized in the energy-requiring reactions of photosynthesis.

Linking Exergonic and Endergonic Reactions

Some activities are obviously energy-*yielding*. The water flowing down a hillside can turn a waterwheel. A match, flaring into a burst of flame burns the fingers. Other activities are equally obvious as *requiring* energy — for example, a stalled car being pushed up into a driveway, a flat tire being pumped up by hand, or a yard full of scattered leaves being raked into a pile. We speak of activities which yield energy as being *spontaneous*; they occur of themselves. Water on the side of a hill doesn't have to be pushed down. There is nothing spontaneous about a load of bricks waiting to be carried up a ladder.

Chemical reactions are similarly classifiable according to their energy circumstances. A burning leaf or a sugar molecule being metabolized are both energy-yielding reactions. Such reactions are termed **exergonic**. An originally energy-rich molecule or group of molecules is converted into energy-poor products by the breaking of chemical bonds. The energy within those bonds is released, or yielded.

However, in a process like photosynthesis, energy-poor molecules such as carbon-dioxide and water are synthesized into energy-rich molecules of sugar. The input of energy, by sunlight, marks photosynthesis as an **endergonic**, or an energy-*requiring* reaction. The bonds of the sugar molecule hold the energy which was put into the reaction.

By *linking* exergonic reactions with endergonic ones, life accomplishes the process which apparently eluded the speaker, who believed that either physicists *or* biologists could be correct but not both. The Second Law doesn't deny the energy necessary to explain life's orderly state. In fact, the Second Law makes it clear that the ordered state of life is earned at the expense of the sun. Universal entropy increase does indeed occur with life tapping into the downhill flow.

This linkage of exergonic reactions driving endergonic ones is so obvious that one wonders why it is seen as a problem. Nobody is amazed when coal is burned (an excrgonic reaction) and the heat which emerges is used to operate an electrical generating station (an endergonic reaction). The linkage of coal burning and electrical power generation is accepted in the same casual manner as is the burning of gasoline (exergonic) to power automobile movement (endergonic). We accept, even if we cannot really envision the mechanism, the linkage between the metabolizing of food (exergonic) with the muscular output of manual labor (endergonic).

We accept as a very natural state of affairs that energy-yielding reactions are what power all reactions which create orderliness out of randomness. From the jumbled pile of bricks to the structured wall, from the randomly colliding oil molecules to the ordered array of synthetic polymers which carpet our homes, clothe our bodies, and fill our waste dumps, we live in a world driven by exergonic reactions linked to endergonic ones. We read constantly that our profligate use of energy to create the local plastic environment is at the expense of the universal one. But deep down we do not believe that our very fabric, our living tissue, our *life* is at one with the Second Law, with entropy increase, with orderliness temporarily gained at the expense of an ultimately inescapable downhill slide to randomness.

"If a burning match is an example of a spontaneous, exergonic reaction, why do you have to strike it?"

Why Do You Have to Strike a Match?

If a burning match is an example of a spontaneous, exergonic reaction, why do you have to *strike* it? The technical term scientists use to identify this requirement is **activation** energy.

Even nuclear bombs have to be *triggered*. The amount of energy that can potentially be released is not the issue. Matches, bullets, tanks of gasoline, and sugar molecules (**reactants**) all contain potential energy which can emerge in an exergonic reaction. But the reactant is in a *stable* condition. A boulder poised on the edge of a cliff may have great potential energy which can be released, but the boulder is at rest. There is an **activation energy barrier** which prevents it from moving. It requires that some energy be expended to start its fall. Once pushed and under way, the reaction will proceed spontaneously. This is true for the match and the sugar molecule as well. Energy-rich chemical bonds exist in both instances, and once the bond-breaking process begins it will proceed spontaneously. But a small amount of activation energy is required to start the process.

The most common form of activation energy is heat; increased thermal agitation ruptures a few bonds and the energy thus released is sufficient to trigger off the entire reaction. This is what occurs in the striking of a typical wooden match. The friction generates sufficient heat to accomplish activation. Just about everyone remembers the chemical experiments which almost always start off with the lighting of the gas burner. The flask is heated to start the reaction.

We've mentioned sugar molecules as sources of potential energy. Since cells utilize sugars as their most immediate energy sources, it is reasonable for us to examine the role of activation energy as it applies to sugars. Glucose is a quite stable molecule. Kept dry, in a container, it breaks down into carbon dioxide and water so slowly that samples on laboratory shelves appear absolutely unchanged for over a century. I said, "appear unchanged." Actually a few molecules have broken down, releasing their energy, but these are too few and too spread out over time to function as triggers. Imagine our boulder at the edge of the cliff. It appears unchanged over the centuries, still poised on the brink. Is it really unchanged? No it is not. Each raindrop which hits the rock dislodges a molecule or two. Energy is released but too little at a time to really affect the boulder's stability.

If we want to increase the rate of bond breaking in a sample of glucose, to increase it sufficiently to overcome

"But life is not really a flame even though philosophers, poets, and scientists like to use the metaphor."

the **activation-energy barrier** which prevents the glucose from beginning its spontaneous exergonic reaction, we can apply heat. This is what we would do in the laboratory. We could play a bunsen burner flame on a pile of glucose until it ignited and then step back and watch the entire pile go up in flames. This is exactly what my daughter did with the pile of leaves: one match was sufficient to unleash the chemical bond energy which erupted in a bonfire. But life is not really a flame, even though philosophers, poets, and scientists like to use the metaphor.

Let's place our sample of glucose in some water, at room temperature, which contains a population of bacteria. We insert a thermometer and observe our bacterial culture over a period of some hours. The temperature barely stirs, the glucose disappears, and the population of bacteria increases. The activation-energy barrier has been overcome without any obvious input of heat. Now we could resort to the notion of a special life force possessed by bacteria, but that approach takes us back through the discredited thinking of vitalism. Is there a way to explain how the glucose was broken down, apparently without the input of the required amount of activation energy?

Imagine a doorbell. The button must be pressed hard enough to make an electrical contact and close the circuit so the bell will ring. There is a spring behind the button. The spring represents an activation-energy barrier. Sufficient force must be applied to overcome the resistance of the spring so as to close the circuit and ring the bell. Imagine a very weak person who doesn't have sufficient strength to overcome the force of the spring and cannot ring the bell. It is conceivable that you could *weaken* the spring, in other words, *reduce* the level of the activation-energy barrier to the point where the person could ring the bell. It turns out that there is a mechanism for reducing the amount of activation energy needed to trigger the breakdown of glucose.

Catalysts and Catalysis

The process of **catalysis** involves the reduction of the activation-energy barrier. A substance which can accomplish this is called a **catalyst**. Suspended beneath our automobiles are catalytic converters which reduce the amount of activation energy necessary in the chemical reactions which convert toxic exhaust products into less deadly products. Cells possess catalysts which similarly reduce the activation-energy barriers for almost every significant chemical reaction associated with life. These are the **enzymes**.

To get an idea of the way in which enzymes function to reduce the activation-energy barrier, think of a large plate-glass window. As long as the glass is perfectly flat, it is quite strong. You may have been momentarily shocked to see someone accidentally bump into a store window, expecting it to shatter, and then were relieved to see the window survive. But imagine the same window lying on the ground with someone lifting one corner of it. The glass begins to bend. There is an almost instinctive realization that the sheet of glass is being subjected to an internal stress and that its stability is seriously threatened. A little too much bend and there will be a shattering release of the tension.

"One bond location may be required to absorb more energy than it can handle, making the entire molecule vulnerable to a chain reaction of successive bond breaking."

Organic molecules are formed of their constituent atoms connected to one another by chemical bonds. If an organic molecule is subjected to stress by being bent or twisted, the energy locked in the bonds shifts from place to place. One bond location may be required to absorb more energy than it can handle, making the entire molecule vulnerable to a chain reaction of successive bond breaking. We have discussed the attractive forces, such as hydrogen bonds, which can form within and between molecules. It is these kinds of forces which come into play in enzymatically catalyzed reactions.

Multiple Point-to-Point Complementarity

The metaphor most frequently used to convey the idea of the relationship between an enzyme and its **substrate** (the molecule with which it interacts) is the lock and key. The key and the lock relate to one another in a **complementary** manner; the key has its small projections and indentations and the lock's pins (which prevent the mechanism from turning) are designed to rise and fall along the length of the key so as to become aligned in a pattern which allows the lock to turn. The idea of complementary parts is that together they form a whole. A jigsaw puzzle is an example of **complementarity** since each of

"The idea of complementary parts is that together they form a whole."

the pieces has places where it must conform to the shape of its neighbors in order to accomplish the entire structure. It is definitely *incorrect* to speak of complementary parts as being *identical* to one another. The parts *complement* one another.

When an enzyme comes into contact with a potential substrate molecule, the degree to which the two molecules complement one another will determine whether or not there will be a catalytic reaction. If their shapes are such that they can come into close proximity at many points along their surfaces, there may be a set of forces established which place a stress at a critical bond within the substrate. Even though no single attractive force between the two molecules is very great, the fact that there are *multiple* points has the effect of concentrating a number of small forces at a critical location. If a key bond breaks, the entire substrate molecule can undergo a massive energy release in an extremely short period of time. Catalyzed reactions occur at very high rates even though the temperature is quite low.

*"One of the qualities of a truly satisfying explanation is its **extendibility**. This ability of an explanation not only to make sense within the original context but to clarify topics which were not originally seen to be related is particularly convincing in science."*

Specificity and the Controlled Flow of Energy

One of the qualities of a truly satisfying explanation is its **extendibility**. This ability of an explanation not only to make sense within the original context but to clarify topics which were not originally seen to be related is particularly convincing in science. Scientific explanations, as François Jacob reminds us, must not only be possible but must accurately account for the actualitiies they purport to explain.

It was obvious that the chemical activities of life were orderly to the degree that made simple mechanistic explanations very hard to believe. Living things not only release energy from foods but they *coordinate* the released energy into heartbeats, nerve impulses, muscle activity, hormone secretion, and all of the remarkable manifestations of life.

It is one thing to accept the lowering of the activation-energy barrier as the explanation for the ability of life to function at very moderate temperatures, but that appears to tell us nothing about the extraordinary capacity of life to channel the energy into the diversity of functions for which it provides the motive force.

Enzymes and their role in catalysis had been under investigation from the mid-19th century, and by the 1920s literally hundreds of different enzymes had been isolated. The numbers grew into the thousands as biochemists became increasingly capable of working with what turned out to be largely protein substances. Why are there so many? Recall that enzymes must have multiple point-to-point complementarity with their substrates in order to function catalytically. Why do you carry so many different keys? Why not one universal key? Wouldn't it make your life simpler? Indeed it would, and also for the thief who would also need only one universal key. The point of all of this is **specificity**.

Each of your keys fits one particular lock. The price you pay for the assurance that nobody else can open your lock is the specificity, the uniqueness of the lock-key complementarity. The reason why there are so many enzymes is that each of them complements only one (or at the most only a related few) substrate molecules. Suddenly we see the extendibility of the explanation. It is certainly true that the primary role of an enzyme is to function as a catalyst, to reduce the activation-energy barrier. But the very way this is accomplished, multiple point-to-point complementarity, provides the explanation for the remarkably coordinated control of energy flow in living organisms. Each substrate requires the presence of a specific enzyme in order for a reaction to occur at life's moderate temperatures. Energy flow is channeled and coordinated by the availability of the appropriate enzymes.

Metabolic Pathways

The word **metabolism** seems almost a holdover from more vitalistic thinking. It means, simply, the sum of all of the chemical reactions occurring in living things. But when we lump all the reactions together under a single word we really gain very little understanding. The presence of the thousands of different enzymes give coherence and meaning to the word.

The remarkable thing about a manufacturing plant is that the assemblage of machines and workers functions in an integrated and purposeful manner. If one machine and its human operators were to be stamping out thousands of wheels which were the wrong size for the light trucks being assembled, we would be very surprised. We anticipate an orderliness to the manufacturing process. It would make no more sense in a body to be producing multiple copies of a hormone which was totally inappropriate for the circumstances.

The chemical processes which constitute metabolism occur in pathways, sequential steps. Typically, A + B -> C + D -> E + F -> etc. Each step is catalyzed by a specific enzyme. At each position where we show an arrow (->) there must be the specifically appropriate enzyme. Without it, the reaction essentially stops. Instead of perceiving of metabolism as a fogbank of chemical mystery, the reality of metabolic pathways gives us an insight into the controlled mechanism by which substances and energy flow in a patterned and controllable manner.

Where Do Enzymes Come From?

We are fast approaching the moment of truth for the mechanistic explanation of life. If it is true that the orderliness of life is due to the presence of the specific enzymes required for the proper flow of particular metabolic pathways, then what accounts for the correct enzymes being in exactly the right location at exactly the right time? The fact that bichemists spent 100 years discovering that there were thousands of enzymes and that they *were* distributed in precise patterned arrays doesn't explain what forces were operating to produce the enzymes and to assure that they were deliverd to the correct locations at the correct times. The challenge faced by the mechanistic explanation is formidable.

In fairy tales there is often a forest, a shadowed and foreboding place filled with mystery and darkness. In the center of the wood there is a place which holds the key to the mystery, a tower or a castle overgrown with vines giving evidence of its long wait for the coming of discovery. Life's tangled forest has its central tower. It did lie deep in the center, hidden, slumbering in isolation. The path to it was not straight, but once the adventurers came upon it, their ingenuity and time would lead them to the center.

CHAPTER TEN

OF PEAS AND FRUIT FLIES

"[A]n exact determination of the laws of heredity will probably work more change in man's outlook on the world, and in his power over nature, than any other advance in natural knowledge that can be clearly foreseen."

— William Bateson

Nothing I know of better serves as a model for the maturation of scientific explanation than the 3000-year long effort to understand life's *continuity.* Like begets like is a useful generalization, but variation in each new generation is puzzling. For most of the three millennia between Greek thought and the closing years of the 19th century, the many explanations which were presented to explain the similarities and differences between parents and offspring failed to be convincing. The nature of the link between generations drained away between the words which had been spun into webs meant to capture its essence.

Terminology Reveals Attitudes and Convictions

A variety of words were used to come to grips with the forming of a new life. The Greek natural philosophers called the process "generation," and this term continued in vogue until the 19th century. A new being was considered to be "generated." The word "made" is a good synonym. We still use this term as in the "generation" of electricity. This is a term which allows for the involvement of a number of causes. The parents might be involved in the generative process but were not always believed to be causally essential. Many creatures, maggots for example, seemed to be generated or made out of decaying matter; others seemed to arise from water or mud. Theories of "generation" allowed for environmental factors such as light, heat, or moisture to play sufficiently causal roles.

Generation also implied a beginning rather than a continuance. Life was seen as starting entirely anew rather than being an extension of prior parental life. Since all kinds of forces could be involved in the generative process, it was believed that the new being was affected by both parental and environmental influences.

"Inheritance" is a term which in the 18th and 19th centuries had a connotation quite different from its modern biological meaning. In a legal sense, inheritance involves the transmission of an estate from one generation to the next. An estate consists of all of the possessions, and no distinction is made between those which the person inherited from his or her ancestors and those which were obtained during the lifetime of the individual involved. By using this model, it was possible to lump together all of the characteristics so, in addition to blue eyes and brown hair, it was believed that a parent could pass on acquired skill at painting or heavily muscled shoulders developed through years of manual labor. We've discussed this subject in Chapter 4 under the heading **Lamarck and Unlimited Mutability**.

It is important to remember that this book is about explanations and the processes by which they were brought into being. All were satisfying for a time, and were ultimately replaced by more satisfying ones. It isn't

"Correctness, in the scientific sense, is a concurrence between the possible explanation and the way Nature actually behaves."

possible for the mind to see the "correct" explanation as if this one has a special sign identifying it from all of the others. Correctness, in the scientific sense, is a concurrence between a possible explanation and the way Nature actually behaves. It took a long time to sort out the candidate explanations. It should be emphasized once again that scientific explanations are never completed. They are closer and closer approximations of the possible with the actual.

The Influence of the Real World

We think of genetics in very personal terms. Our concerns as to the possibility of our children inheriting a congenital disorder or our own susceptibility to heart disease or diabetes make us very conscious of inherited biological factors. What we forget, in thinking of much earlier times in human existence, is that our ancestors had no science called genetics to which to turn for answers to questions of "Why me?" Karma, luck, the evil eye, fate, the bad seed, God's will, maternal influence, and the sins of the father visited on the children were some of the answers given to such questions.

It was known that hair coloring or lip shape or extra fingers and toes "ran in families," but the patterns were unclear. So long as it was believed that the environment and hidden forces played a major role in the outcome of human development it was highly unlikely that much progress in biological inheritance would be made. In spite of the obstacles, however, in one area of human endeavor there was a concerted and long-standing effort to bring order out of chaos.

The study of "inheritance" was a very important issue in agriculture. Breeding was at the core of a successful agricultural society. Desired traits and their transmission to offspring were critical. At the time inheritance was modeled on the legal concept of the passing on of an estate, a horse was seen as being capable of transmitting not only what we would call its innate qualities but also the influence of the environment in which it had lived its life. Thoroughbred breeders in 18th-century England insisted on importing stallions from Arabia because they were convinced that it was the adaptation to the desert environment which gave the Arabian breed its speed and stamina. They were absolutely certain that an Arabian horse born and raised in England would have "degenerated" and would have lost some of its desirable qualities.

It was this blending of what we would consider truly heritable qualities with acquired ones which made the studies conducted by animal and plant breeders so confusing. In addition, consider the fact that thoroughbred "blood lines" were traced only through the male parent. The contribution of the mare was considered "environmental" and fell into the same category as ample and wholesome food and water.

The Concept of Biological Continuity

There were several theories which stressed that life was not generated anew with each generation but was a continuation of parental life. There were suggestions as to how this continuitive process was accomplished. In the 18th century, for example, it had been suggested, by Buffon, that each organ of the body produced particles (seeds) which contained the essence of that organ. Particles from the liver, the lungs, the brain, and so forth were thought to be sent by way of the blood to the reproductive organs where they were incorporated into the reproductive fluids. The mixing of male and female fluids at conception resulted in a combining of these essential particles. In the 19th century, Charles Darwin resurrected the theory under the name "pangenesis." Darwin accepted the possibility that the environment could modify or affect the parental "pangenes," and thereby he incorporated Lamarckian thought into his concept. You may recall that Buffon resorted to the idea of an *internal mold* as the device which would guide the particles into the proper relationships so as to give the body its proper form.

"Darwin's pangenesis ideas never were well received. But the idea of particles rather than forces was to have a major role in shaping our ideas about what is involved in establishing the continuity between generations."

Even though "particles" sounds a lot better than vague "forces" in a mechanistic explanation, reliance on explanations which require internal molds made most 18th- and 19th-century scientists very uneasy. Darwin's pangenesis ideas never were well received. But the idea of *particles* rather than *forces* was to have a major role in shaping our ideas about what is involved in establishing the continuity between generations.

Gregor Mendel (1822-1884)

It isn't easy to properly describe this Austrian monk. The undisputed facts are that he was born into a peasant family, entered a monastery in Brno, Czechoslovakia, attended the University of Vienna for two years, failed the examination which would have certified him as a teacher, returned to the monastery where he did his research and ultimately became abbot. As to what kind of a person Mendel was, the reasons that he performed his investigations in the manner he did and the theoretical concepts he had in mind have been argued by geneticists and historians of science for many years without any consensus emerging. He has been portrayed by some as a very simple-minded and extraordinarily lucky person and by others as a genius with truly astonishing insight. While Mendel the man remains obscure his investigations have the clarity and decisiveness which rank them among the most significant in the history of science.

The province of Moravia in which Mendel lived and worked was a center of practical animal and plant breeding. Mendel was doing what many of his fellow Moravians were doing — **hybridizing**. The crossbreeding of pure parental lines was a vigorously pursued practice in the effort to develop organisms which combined desirable parental traits, and, above all, to discover the rules by which inheritance operated. There was a concerted effort to reduce the guesswork and failure rate which accompanied the efforts to increase agricultural productivity. In part, therefore, Mendel was a very practical man, a product of his time and place.

Something else was involved. Linnaeus had claimed that entirely new *species* could be brought into existence by the hybridization of existent ones. This was an idea which attracted a great deal of attention, and prizes were offered by various scientific academies for answers to the basic question as to whether or not **hybrid** creatures could be the founders of entirely new species. Mendel seems to have been fascinated by the evolutionary implications of inheritance. In this regard Mendel goes beyond the practical to the deepest meanings of biological speculation.

It was well known that the *offspring* of hybrids often "reverted" to the parental traits, and Mendel seems to have been pursuing the details of this process as he traced several generations of offspring in various mating combinations. It was the reappearance of the original parental traits, *unadulterated and unchanged*, after their disappearance for several generations which appears to have revealed to Mendel the key to the problem. As we have previously mentioned, biologists including Buffon and Darwin had speculated that inheritance involved the transmission of particles of matter. As a result of the outcomes of his breeding investigations, Mendel provided indirect but very convincing evidence that inheritance was indeed accomplished by *particles*, actual physical pieces of matter, which retained their basic natures no matter how many times they were passed to offspring and no matter what environment they encountered. This conviction, that inheritance was accomplished by discrete particles which resisted change, permitted Mendel not only to provide an explanation for the outcomes of the breeding experiments but more significantly, to accurately *predict* the outcome of future breedings. The vision of a rational agriculture was attainable, and beyond it was the possibility of the mechanism of life's continuity.

"The vision of a rational agriculture was attainable, and beyond it was the possibility of the mechanism of life's continuity."

The Use of Probability in Mendelian Thought

Although there are some specific instances in which a geneticist can state flatly that a child will inherit a certain trait, most genetic predictions are phrased in probabilities. We are frustrated when we are told that there is a 50:50 chance that some inherited trait will appear in a child. We demand a higher level of predictive precision. After all, we say, genetics is supposed to be a *science*. The implication of this attitude is that true science is absolute. Balls released from the hand *always* fall down. There is no 50:50 evasion of the laws of gravity, but Mendel's laws are full of probability statements.

Let's ask a physicist, using the immutable laws of Nature, to predict the result of a single coin toss — heads or tails. There is no frustration when the physicist informs us that the *probability* is 50:50 that the toss will come up heads. It is similarly 50:50 that it will come up tails. Why do we accept and understand the uncertainty of coin tossing and feel so uneasy and unsatisfied with probability predictions from biologists?

"We think of biology as a fuzzy science which someday may achieve the assurance of physics. In this particular case, the predictive accuracy of the coin toss and the inheritance outcome will never get less fuzzy. That's because they are based upon the same process of randomness."

I suspect that, among other things, the explanation involves differing levels of confidence in the knowledge of physicists and biologists. We think of biology as a fuzzy science which someday may achieve the assurance of physics. In this case, the predictive accuracy of the coin toss and the inheritance outcome will never get less fuzzy because they are based upon the same process of randomness.

There are two sides to the coin. The outcome of an honest, unbiased toss can only be stated in probability terms. Mendel's particles behave as honest, unbiased alternatives. When there is a genetic situation where the particles can combine in either of two ways — just like the head and tail outcome — it will *never* be possible to predict, in a given case, which of the two will occur. This is not a case of incomplete knowledge.

Big A and Little a; or Why Use Garden Peas?

There are very few college students who have not been exposed to the terminology and symbols used in **transmission genetics**. The term "transmission" refers to the manner in which the genetic elements or units are passed from parents to offspring — the inheritance "rules of the road." There are other genetic topics (e.g., recombinant genetics, molecular genetics) which we will consider separately. But the basic patterns by which genetic material is passed from generation to generation is what most of us associate with the diagrams involving "big A and little a." It was the insight of Gregor Mendel which provided us with this penetrating view beneath life's surface exterior.

The selection of the organism to use in a study (the biological **model**) reveals a great deal about the mind-set of the investigator. In Mendel's case, his use of the garden pea (*Pisum sativum*) tells us that he was intent upon hybridizing parental lines which had very distinct differences. For example, one parent came from a so-called **pure-breeding** stock in which all the individual plants were *tall*. The other parent was selected from another pure-breeding stock which was *short*. By following the height of the offspring in several successive generations Mendel hoped to clarify the basic transmission patterns. In all cases he chose traits with *distinct* *alternatives*: yellow seeds or green seeds, round seed shape or wrinkled seed shape, purple flowers or white flowers. This approach begins to explain the "Big A, little a" notation which we all remember from our first exposure to genetics. The upper case letter (A) represents one alternative and the lower case letter (a) represents the other. Which trait goes with which letter is decided *after* the outcome of a hybridization experiment is known. The upper-case letter is assigned to the trait which turns out to be dominant and the lower case letter to the trait which turns out to be recessive. We'll discuss the full implication of these terms shortly.

It may sound like intellectual cheating to believe that science should make up its mind *before* it carries out an experiment. Once you accept the fact that an experiment is a device to *decide* between possible alternatives, you see that it has to be this way.

What Shall We Call the Particles?

Today we call the units of genetic transmission **genes**. Mendel referred to the particles as "factors," or in other words, "things." He did not give them a functionally descriptive name. He didn't know enough about his supposed factors or specifically, he didn't know where they were located, what their structure was, how they executed their roles. The naming process in science is often very contentious. We'll return to this topic later.

For our puposes it will be prudent to use the current terminology. We speak of the **gene** for plant height, for example. If that gene can exist in two alternative forms (tall and short) we use the term **allele** when we refer to the alternatives. Some genes exist in several alternative (allelic) forms.

Mendel first reported the results of his studies in 1865 before a relatively small group of people at a meeting of the Brno Natural History Society. A year later a written version was published and circulated widely throughout Europe. His work apparently escaped the attention of scientists until 1900 when three researchers, DeVries (Holland), Correns (Germany), and Tschermak (Austria) each independently rediscovered the basic Mendelian concepts and then found the publication which had remained unrecognized for 34 years.

As to a name for the particles, several different ones were used until 1909 when Wilhelm Johannsen suggested "gene." This one was adopted, and we will use it in our discussion of Mendel's work even though the term didn't appear until 25 years after Gregor Mendel's death.

Mendel's First Law: Segregation

P AA x aa

The parental (P) generation.

To segregate means to separate. Mendel's system of inheritance assumed that each trait (let's use height of the plant as an example) was controlled by two alleles: one allele inherited from each parent. We will use the upper case **A** to designate the allele for *tall* and the lower case **a** for *short*. The parental generation is symbolized as **P**.

In this case one parent possesses two alleles for *tall* and the other parent possesses two alleles for *short*. You may have noticed that I have not designated one parent as being the male and the other as being the female. For most "garden variety" genetic problems this is a good idea. One of Gregor Mendel's insights was that alleles would not behave differently in the body of a male than in the body of a female. In general this is true, but we have recently learned that there are some fascinating exceptions to the general rule.

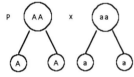

In the act of segregation, only one allele from each parental pair is placed in a future reproductive cell.

In Mendel's thinking, each parent can contribute only *one* allele to an offspring; therefore there must be a **segregating** of each parent's alleles. As a result of the process of segregation each allele can behave as an individual unit. It has become disconnected from and can flow into the next generation without its partner.

The Creation of the F_1 Generation

In sexual reproduction, the two parents contribute genes to the next generation. Each member of the **first filial** (F_1) generation (filial = pertaining to a son or daughter, namely, offspring) will possess one **A** allele and one **a** allele as the result of the union of reproductive cells from the two parents. Mendel was very conscious of the possible alternatives at this step, and he correctly identified the role of random chance in the coming together of the parental alleles. Either one of the **A** alleles can combine with either one of the **a** alleles. There are four

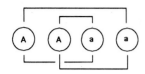

The four ways to combine the parental alleles.

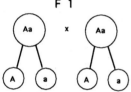

F 1

Segregation of gametes in a cross between two members of the F_1 generation.

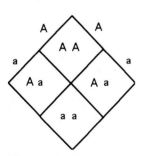

The Punnett Square display of the possible F_2 combinations.

different ways to reach the common outcome; all of the F_1 offspring will have the same allelic pattern, namely, **Aa**.

It is clear that every member of the F_1 generation will have received an **A** allele from one parent and an **a** allele from the other. There is just no other way for the alleles to be transmitted. So the *genetic content* of the F_1 generation is beyond question. What remains to be determined is their appearance.

Mendel's records revealed that every member of the F_1 generation was *tall*. The presence of the **A** allele had completely dominated the expression of the **a** allele. There were absolutely no *short* plants at all. It became clear from the crosses he performed that some alleles were **dominant** while others were **recessive.** The concept of dominance and recessiveness becomes clear only by actually breeding individuals with the two alternative traits. Only by observing which trait disappears in the F_1 generation can we discover which is the dominant and which is the recessive trait. We designate dominant alleles by using the upper case and recessive alleles by using the lower case letter. In this F_1 generation, the recessive *short* trait has completely disappeared.

The Reappearance of Recessive Traits in the F_2 Generation

If we cross two members of the F_1 generation, following the rule of segregation, we separate the alleles as indicated. Now when we combine alleles from each parent, we find that once again there are four ways to do so. If we use the **A** from each parent, we get offspring with **AA**. Similarly, if we use the **a** from each parent we get offspring that are **aa**. There are *two* different ways to combine an **A** from one parent with an **a** from the other. The possible combinations for the F_2 generation are **AA, Aa, Aa, aa**. You may recall using a **Punnett Square** to help you keep things straight.

Genotype and Phenotype

The same man who coined the term "gene," Wilhelm Johannsen, also formalized the distinction between the genetic content, called the **genotype** (for example, **AA**), and the trait which resulted from the expression of the genes, called the **phenotype** (in this case, *tall)*. The importance of this distinction is apparent when we examine an individual with the genotype **Aa**. Since such an individual possesses two different forms of the gene, what will be the *expression* of the genes? In the specific case which Mendel examined, height of the plant, it turned out that an individual with the genotype **Aa** was *tall*. The presence of the **a** gene was completely masked. It is evident that there can be *two* genotypes which will produce an individual with the phenotype *tall*: **AA** and **Aa**. The appearance of the individuals is identical. In such cases the **A** gene is completely **dominant**. The **a** gene is said to be **recessive**.

"It was the simple arithmetic involved in the Aa×Aa segregation which explained the 3:1 ratio."

The most striking feature of the F_2 generation is the *reappearance* of the *short* trait. It had disappeared completely in the F_1 generation. The proportion or ratio is three *tall* to one *short*. The reappearance of the *short* trait in *predictable ratios* gave Mendel two very important insights. The first was that the **a** allele does indeed behave as a discrete particle. Even though it is masked in the F_1 generation, it is not modified by its environment. It reappears totally unchanged in the F_2 generation. The second insight traces from the *proportion* 3:1 of *tall* and *short* individuals. This ratio, appearing in experiment after experiment, meant that there was a precise mechanism involved. It was the simple arithmetic involved in the **Aa   Aa** segregation which *explained* the 3:1 ratio.

Are All Alleles Either Dominant or Recessive?

The best way to respond to the meaning of dominance and recessiveness is to state what a gene actually does. As is detailed in Chapter Twelve, a gene is a chemical which causes certain chemical reactions (—>) to occur; and in the simplest possible terms, the dominant allele **A** —> product *A* and the recessive allele **a** —> product *a*. It is the *interaction* of the chemical products *A* and *a* which determines the phenotypic expression. In the case of the pea plants, product *A* whether present in a "double-dose" (**AA** —> *AA*), or a "single-dose" (**Aa** —> *Aa*) is sufficient to produce a *tall* plant. Only when the genotype **aa** is present is **no** product *A* formed and the plant is *short*. It is possible for a genetic circumstance to exist in which the two products interact to produce an outcome which is *intermediate* between the two extremes. Let's create a case in which the product of gene **B** produces a red pigment: **BB** —> product *BB* = red. The product of gene **b** produces a white pigment. If the genotype is **bb** then the product would be *bb* and the phenotype would be white. But if the genotype is **Bb** —> product *Bb* then *both* red and white pigments are produced. An individual with the genotype Bb would be pink, intermediate between red and white. In such cases, the genes are spoken of as showing **incomplete dominance**.

Two Terms Which Will Be Helpful

One of the problems with explaining genetics is the terminology. Things really bog down when the names of parts or processes are injected into a conversation. Incidentally, the terminology is not the only problem; genetics is, for many people, not very *sensible*. The use of symbols (Aa) to represent things removes genetics from the senses. Instead of seeing real plants or animals, symbolic representations have to be manipulated inside the brain. There is a genuine need to do things this way, but the use of symbolic representation does produce a feeling of unease in many of us. Genetics is like mathematics in this regard.

As we have seen, a *tall* (that's its *phenotype*) pea plant can have either of two *genotypes*. It may be **AA** or **Aa**. It is important to know which. The term **homozygous** is used to describe the condition in which both alleles are identical (**AA**), and the term **heterozygous** is used to describe the condition in which they are not (**Aa**). Of course the condition (**aa**) is also homozygous. So it is essential to describe **AA** as being **homozygous-dominant** while **aa** is termed **homozygous-recessive**. The heterozygous condition (**Aa**), since it consists of one dominant and one recessive allele, needs no further clarifying designations. Geneticists use these terms constantly, as in a conversation when they say, "The male parent is homozygous recessive for the trait in question (**aa**) and the female parent is heterozygous for the same trait (**Aa**). Therefore it is equally probable for them to have offspring with either the heterozygous or homozygous-recessive genotypes." Try a Punnett square to see if you follow the reasoning.

Mendel's Second Law: Independent Assortment

If genes are really as Mendel conceived of them, totally discrete particles, then they should behave in predictable ways. Imagine a box in which there are two pairs of marbles; one pair of marbles is white and the other pair is gray. One white marble has an **A** written on it, and the other white marble has an **a** on it. One gray marble is marked **B,** and the other gray marble is marked **b**.

If you were asked to put *one* marble from each pair into a set of smaller boxes, there are four different combinations you could form. As long as the marbles can truly be separated from one another all of the possible combinations can be formed. Gregor Mendel's **Law of Independent Assortment** is based upon the conviction that the hereditary units, the genes, are actually individual particles and will behave like individual particles.

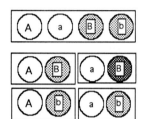

One white and one gray marble from the large box can be distributed in four different combinations into the smaller boxes.

For the traits with which Gregor Mendel worked, the Law of Independent Assortment held true. The four smaller boxes in our diagram represent the kinds of reproductive cells which a parent (the large box) can form. A reproductive cell is called a **gamete**; the female gamete is an **egg**, and the male gamete is a **sperm**. The Law of Independent Assortment begins to make clear one of life's most puzzling qualities, **variation**. Examine the large box. This particular individual is heterozygous for the two traits controlled by the genes **Aa** and **Bb**. If the **A** and **B** genes are completely dominant, this individual would display the dominant phenotype for both traits. But observe that the gametes this individual can form include one alternative (the box on the lower right) in which the two **recessive** genes have been combined. The potential for variation in the next generation is made clear in such a system. When we realize that the other parent is similarly shuffling his/her genetic content in forming gametes, we get a glimpse of the potential for recombining parental genes in a variety of patterns.

In our example we have used the simplest possible case, only two sets of genes assorting independently. Since most creatures have many thousands of pairs of genes, the possibilities for producing various gene combinations in the gametes is astounding. The next time you are at a family dinner with your biological parents and your brothers and sisters, reflect for a moment upon the ways in which individual genes have been shuffled and recombined to produce the variations in a single family.

"In science, as in all human endeavors, linearity is rarely the pattern. It is important for us to deal with the many-tracked way in which science actually develops its explanations."

Elements and Chromosomes

Throughout this book the story of biological explanation has been told in a roughly *linear* fashion. By this I mean that one event follows another in more or less a straight story line. This simplistic style makes following the events easier, but it distorts historic reality. In science, as in all human endeavors, linearity is rarely the pattern. It is important for us to deal with the many-tracked way in which science actually develops its explanations.

I have previously compared scientific theories with webs. In Chapter One it was pointed out that individual strands of thought connect with one another so as to form an ever enlarging structure, and it is the fact that individual pieces *do* fit coherently into the developing pattern that gives science the assurance that its constructs are valid representations of reality.

In the year 1873, three biologists independently observed and recorded what we would eventually call **chromosomes**. A. Schneider, Otto Butschli, and Hermann Fol are not household names recognized instantly like Gregor Mendel. They were studying the structure of cells and were particularly interested in the way new cells came into being. Schneider drew remarkably accurate pictures of chromosomes, drawings immediately recognizable as portraying their replication in **mitosis**. Oddly, it seems to us now, he made no connection between the behavior of the little strands of nuclear material and the inheritance of genetic traits. We need to be reminded that Gregor Mendel's theorizing was not rediscovered until 1900. The little strands were just that, little strands. The cell is full of all sorts of structures, and *cytologists* (cyto = cell) were faced with the task of providing meaning and coherence for the wealth of material the microscope was revealing to them. It is exactly the same problem which faces workers in an archeological dig. All sorts of previously unknown scattered fragments which the diggers uncover must be given meaning. In both cases it is the conviction that what has been uncovered *does* have meaning and that the human intellect can imagine that meaning which drives the investigators.

By 1875, Edouard Strasburger had drawn attention to the universal presence of chromosomes in the cell nucleus and focused the attention of his fellow scientists upon these tiny strands of material which stained so intensely (chromo = color). Four years later Walther Flemming was able to provide detailed drawings of the chromosomes in the various stages of cell division and, most importantly, showed that during the process each chromosome divided into two equivalent "daughter" chromosomes. In that same year, Oskar Hertwig showed that fertilization involved the fusion of two parental nuclei with the bringing together of their contained chromosomes. All of these observations had occurred without any of the investigators having had the slightest inkling of Mendel's discovery of the basic rules of transmission genetics.

"Sutton saw in the behavior of chromosomes the same pattern that Gregor Mendel had proposed for his 'elements.'"

With the rediscovery of Mendelian genetics in 1900, it was only a matter of time before the similarity between the postulated behavior of Mendel's particlelike factors and the behavior of the visible chromosomes would become evident. Walter S. Sutton was a graduate student at Columbia University. His advisor, Edmund B. Wilson was the foremost American cytologist. Sutton had been given a project involving the chromosomes of a grasshopper of the genus *Brachystola*. The reason for using this creature is very informative as to the importance of using an appropriate biological model. In many organisms all of the chromosomes are about the same size and shape and it is very difficult to distinguish exactly what is going on. But in the insect used by Sutton the chromosomes are all quite different from one another and he was able to see that they existed in distinguishable pairs. He was able to follow individual pairs of chromosomes through the cell cycle. Sutton saw in the behavior of chromosomes the same pattern that Gregor Mendel had proposed for his "elements." There was **segregation** of the chromosome pairs so that only one member of a pair entered each gamete. A parent's **Aa** chromosome content was segregated with the **A** gene going to one gamete and the **a** gene going to another.

Sutton counted the chromosomes in the gametes and found the number to be exactly half of the number in a typical body cell. Sutton called this a "reducing division" for obvious reasons. If the body cells contained 22 chromosomes (11 pairs), the gametes contained only 11 chromosomes, one member from each pair. Finally, when two gametes fused at fertilization, one egg with one sperm, Sutton saw that the original chromosomal number was restored — 11 chromosomes having been donated by the egg and 11 chromosomes by the sperm. Not only was the number 22 restored, but the *pairing* of the chromosomes was also visibly confirmed. In 1903, as a result of his observations, Sutton was able to speculate:

> I may finally call attention to the probability that the association of paternal and maternal chromosomes in pairs and their subsequent separation during the reducing division ... may constitute the physical basis of the Mendelian law of heredity.

With this single sentence, Sutton brought together two parts of the web, and they fit one another precisely. Mendel had imagined genetic particles in 1866, and the cytologists had seen chromosomes since 1873. The graduate student who was using appropriate biological material, the reproductive organs of a grasshopper, was able to join the theoretical with the actual.

Mendel's Second Law in Trouble

Almost as if it was a soap opera, the genetic story was not meant to have a tranquil history. While chromosomes provided a physical basis for Mendel's First Law, Segregation, Sutton pointed out that they would cause a serious problem for the Second Law. Surely a grasshopper had more than 11 pairs of genes. With eyes, legs, wings, antennae, inner organs, colors, textures, and so on it was inconceivable that a chromosome could *be* a gene. There just weren't enough chromosomes for each of them to be or contain only one gene. But *independent* assortment required that each gene be totally disconnected from all other genes so as to produce the scrambling and shuffling we discussed using the marbles in the boxes model. This theoretical problem was shown to be a reality in 1906 by William Bateson and R. C. Punnett. Each chromosome contained many, many genes. All of the genes present on

*"All of the genes present on one chromosome are said to be **linked**. Like the people in a boat, wherever the boat goes so do the passengers."*

one chromosome are said to be **linked**. Like the people in a boat, wherever the boat goes so do the passengers. Mendel was essentially correct in that each gene retains its integrity as it passes through the generations, but he had not foreseen that *groups* of genes (**linkage groups**) would travel together. *Chromosomes* assort independently, but the genes they contain are not as free.

Why the Fruit Fly?

In another laboratory at Columbia University, in 1910 Thomas Hunt Morgan (1866-1945) and his graduate students Alfred H. Sturtevant (1891-1970), Calvin Bridges (1889-1938), and Herman J. Muller (1890-1967) began a series of studies which, five years later, would result in the publication of a book that would bring Mendelian genetics into full maturity. The publication of *The Mechanism of Mendelian Heredity* in 1915 made the fruit fly, *Drosophila melanogaster* synonymous with genetic experimentation.

Morgan had started his scientific career as an embryologist and would write (in 1927) a textbook in embryology which was to become a classic. But the nature of inheritance captured Morgan's passion even as the problems of embryonic development continued to haunt him. One of the problems the Morgan group investigated was the sudden and unanticipated appearance of an inheritable trait — a mutation.

The Dutch botanist, Hugo de Vries, one of the three rediscoverers of Mendel's concepts, was the first person to suggest that the alternative forms of a gene had evolved by the process of **mutation**. In de Vries mind, an original gene **A** would have changed (mutated) into the alternative form **a** in some ancestral individual. Although we now know that these random and unanticipated changes are due to modifications in the chemical nature of the gene, the only thing known in the early twentieth century was that once changed, the mutant gene would flow through the generations following the same transmission rules as its unmutated alternative. T. H. Morgan attempted to put this idea to experimental verification. To do so he realized that he would have to raise enormous numbers of creatures since, in theory, mutations were not very common. If he was fortunate enough to discover a

"The idea was tantalizing since it was a way to account for the new traits which powered evolutionary variation. Lots of things were coming to a head. Biology was approaching its maturity."

mutant individual, Morgan also realized that the hoped-for mutant would have to be fertile because it would be necessary to breed it in the hope of tracing the fate of the mutant gene through many generations to verify its continued existence. The idea was tantalizing since it was a way to account for the new traits which powered evolutionary variation. Lots of things were coming to a head. Biology was approaching its maturity.

William E. Castle of Harvard University had suggested to Morgan that *Drosophila melanogaster* was an excellent organism for genetic research. The rapidly maturing and prolific fruit fly had been found to prosper in the laboratory. It could be reared in glass pint bottles (which at that time were used commercially as milk and cream containers and were available in the thousands), and millions of individuals could be bred, reared, and studied in relatively small and simple laboratories. As evidence of the suitability of *Drosophila* and the techniques of Morgan's group, exactly the same procedures instituted at Columbia continue to be employed worldwide at the forefront of genetic investigations.

Drosophila found in the wild, either out of doors or drawn to a lunch-time banana, have red eyes. The first mutant Morgan's group identified was a white-eyed male. This single insect was to yield progeny which the Columbia geneticists would use to start the field of study familiarly called *Drosophila* genetics. It would lead to the Nobel prize, to the mutational effects of ionizing radiation, and to the medical and legal implications based upon the predictive power of genetic theory. The genetic basis of the determination of sex and the roles of genes in orchestrating embryonic development all trace to the little laboratory, the "fly room," at Columbia University.

When Is a Linkage Group Not a Linkage Group?

A soap opera depends upon twists and turns of plot to keep its audience tuned in. As we have seen, Mendel was absolutely correct about segregation but only partially correct about independent assortment. When Bateson and Punnett verified that all the genes located on a given chromosome would travel together as a linked group, it was thought that the transmission rules of the road had been learned. If genes **A** and **B** were located on the same chromosome then **A** and **B** would have to go to the same gamete. After all, wherever the boat goes, so do the passengers.

"This is a particularly good example of the kind of circumstance which leads many students to throw up their hands in disgust and demand of a teacher, 'Which of these rules do you want us to know?'"

Morgan's group found, to their surprise, that although linked genes (all the genes located on a given chromosome) did indeed travel together as a general rule, sometimes they did *not*! In some cases, even though **A** and **B** started out on the same chromosome, they became *unlinked* in the process of gamete formation. It was learned that in a predictable percentage of gametes, **A** and **B** had somehow broken the linkage which should have united them. **A** had ended up in one gamete, and **B** was in an entirely different one.

This is a particularly good example of the kind of circumstance which leads many students to throw up their hands in disgust and demand of a teacher, "Which of these rules do you want us to know? Should we believe Mendel and the independent assortment of his Second Law or should we believe Bateson and his linkage which violates the independent assortment of the Second Law or should we believe Morgan who demonstrates that it is possible to break linkage?" Students communicate an attitude which states that when science has made up its mind who is right, then and only then will they mentally return to be given the final and correct answer. Not only is science an ongoing process, but it can be a combative one as well. There are often strong convictions held by intellectually aggressive persons.

What Is the Physical Basis of Linkage?

Before you respond to this question by saying, "Who cares?" let me suggest that you really do but may not realize why. Most people are aware that we have, today, techniques which permit us to identify the location, on the chromosome, of the genes for a variety of human traits and disorders. Not a week goes by without a newspaper article announcing that the gene for some condition or other has been identified. Every one of these "breakthroughs" is dependent upon the work done by the Morgan group, the work which established the structural features of the chromosome.

I used the metaphor of people in a boat being obliged to go where the boat went. But that doesn't tell us how the people were arranged in the boat. Did everyone have an assigned seat? Were they seated side by side in rows? Were they free to stroll around?

Morgan's research group began accumulating data about genes which were linked. In theory, all genes located on a given chromosome should always remain linked. In actuality, however, the results of thousands of matings showed that linkage was sometimes broken. Careful record keeping of the frequency of such breaks revealed that genes which were normally linked lost their linkage with very definite frequencies. Some genes were found to be "tightly" linked (that is, they became unlinked extremely rarely) whereas other genes lost their linkage more frequently and were described by the group as being "loosely" linked.

"But how does one deal with data which show that linkage is the general rule but can be violated with varying degrees of frequency?"

I suspect that most of us would have responded with despair to the kind of data Morgan's group was getting. The concept of a chromosome as a definite structure carrying a specific group of genes makes sense if the genes always travel as a linkage group. However, if genes travel as individuals, not being carried on structured chromosomes, then totally independent assortment makes sense. But how does one deal with data which show that linkage is the general rule but can be violated with varying degrees of frequency?

You may recall that Sutton had brought together the theoretical Mendelian particle and the actual, visible chromosome. It is important, in considering what happened next, to keep in mind the developing understanding of the cell's nucleus and its contained chromosomes. **Cytology** had been providing increasingly convincing evidence that chromosomes were the visible structures which passed from parents to offspring and that these chromosomes did so in precisely patterned movements. Any consideration of genetic theory had to take into account and be in accord with the structure and function of the chromosome.

T. H. Morgan's Conversion

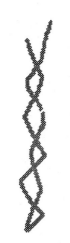

Janssens'representation of chromosomal chiasmata.

Prior to 1910, when he was still deeply involved in the problems of embryological development, T. H. Morgan was highly skeptical of the physical reality of the gene and the significance of the chromosomes. It comes as a surprise to most people that the man who was to place genetics upon such a sound basis originally held the idea of the gene in such low esteem. He was concerned that biologists were accepting, in the concept of the gene, a preformed little package which claimed to solve very complex embryological problems. Today we use the term "black box" to describe such concepts: an explanatory process or device without any evidence of just how it works. Morgan seems to have had a particularly strong aversion to ideas which had a black-box quality about them.

It wasn't until he became intimately involved in the *Drosophila* studies that he began thinking of genes as truly existent and of chromosomes as critically important in the process of genetic transmission. But what kind of a gene container was the chromosome that it could yield such confusing results? Morgan was aided by some observations published in 1909 by F. A. Janssens who had seen chromosomes which appeared to be twisted around one another. This gave Morgan an idea as to how genes might be arranged on the chromosome to account for the linkage data he had obtained.

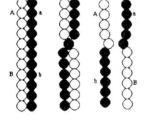

The beads-on-a-string model of the chromosome showing crossover recombination.

From Chiasmata to Crossing Over

Janssens' observation led Morgan's group to propose an idea which was frequently called the "beads-on-a-string" model. The twisted crossings (chiasmata) suggested that paired chromosomes might sometimes *exchange* segments resulting in new genetic combinations. Morgan's concept as to the *linear* arrangement of genes on the chromosome explained the varying *degrees* of linkage his data had shown. If chromosomes could cross over and reform so as to have new combinations of genes, it stood to reason that the farther apart two genes were located on a chromosome the more opportunities existed for breakage between them to occur.

It was Sturtevant who pursued the idea that it would be possible to **map** the chromosome. All one had to do was keep records of the number of **crossover recombinations** which occurred between genes which should have remained linked. The higher the percentage of recombinations, the greater the distance between the genes being studied. The actual work was heroic. Many years of studying crossover occurrences between dozens of linked genes yielded ever more precise **crossover-frequency maps** of *Drosophila*'s four pairs of chromosomes. Each gene was assigned a **locus** (location), and its various alternative (allelic) conditions (for example, **A** or **a**) were normally found in that spot on the chromosome.

Mirroring Morgan's slow acceptance of the reality of the gene, William Bateson was for a very long time unconvinced that the genes were arranged in any limiting manner on the chromosome. This was the man who had so vigorously championed Mendel and had discovered linkage. Scientists, like other mortals, have to struggle against strongly held preconceptions.

What's in a Name?

Earlier in this chapter I explained that we would refer to Mendel's particles as genes even though that name was not introduced until 1909 by Johannsen. I also indicated that we would consider the nature of naming things and the discord such activities can generate. This may seem like a petty issue; after all, as Shakespeare pointed out, a rose by any other name would smell as sweet. But there is something else at stake here and we would be well advised to understand it.

Shortly after the rediscovery of Mendel's work in 1900, the scientists who pursued genetic studies began seriously debating what they thought the "particles" were and did. One of the early suggestions was made by William Bateson. This British geneticist has been largely forgotten and is almost never mentioned in biology textbooks. This is odd since he was the man who suggested the name **genetics** for the subject we are discussing. In 1902 Bateson wrote a book, *A Defence of Mendel's Principles of Heredity*, which brought Mendelian thought up to date by placing it within the context of the cell theory. It was in this book that Bateson introduced such terms as "homozygous" and "heterozygous" to clarify the genotypes. In attempting to get at the relationship between genes and traits, Bateson used the term **unit-characters**. The hyphen (-) separated two aspects of his genetic thinking. The **unit** referred to the *particle* itself while the term **character** was a reference to the *trait* which was produced. To give you some idea of why this topic was to become very contentious, think about what you are saying when you tell someone that you have inherited your father's brown eyes. If you conceive of a particle which is a "genetic" brown eye and in some unknown manner brings into existence a "real" brown eye you can understand the meaning of the term "unit-character." Bateson used the hyphen to indicate that the particle was certainly *connected* with the trait but wasn't the trait itself. But most importantly for an understanding of the debate which was to come, the *unit* was irreversibly connected with the *character*. If you had one, you were going to get the other. Unit-characters are absolute; if you inherit a unit you will express its character.

*"When Johanssen suggested the name **gene**, he wasn't arguing with Shakespeare. He was arguing with Bateson and the other scientists who imagined the particles as being* equivalent *to the trait."*

But suppose you have an entirely different idea of what this particle is. For example, suppose the particles function like the ingredients in a cake recipe. The eggs are not a cake, the milk is not a cake, nor is the flour or the oven a cake. This is an important distinction. The individual ingredients interact with one another in a very complex manner. No one of them is the cake. Depending upon the temperature of the oven and the way the ingredients were combined, you may get cookies, crackers, or pancakes.

When Johanssen suggested the name **gene**, he wasn't arguing with Shakespeare. He was arguing with Bateson and the other scientists who imagined the particles as being *equivalent* to the trait. He wanted to establish a name which communicated the belief that the particles were involved in the ultimate production of the trait but were *not* the trait itself.

If one thinks of a gene as the *equivalent* of a trait (in other words, as a unit-character), then it follows that the absence of the "cake" gene means that some unfortunate person will have to go without dessert. There are those to whom this is an intolerable thought. They therefore reject the idea that the gene is significantly involved, particularly in critical human traits such as intelligence or personality. They insist that the provision of an adequate environment can compensate for any genetic deprivation. They argue that if you cannot bake your own cake the environment can provide one from the bakery.

But the gene is *not* a unit-character. It is one of a number of functioning components in a complex set of interactions. Together with many other genes and the products of other gene actions, as well as a host of environmental factors, each gene represents a *fragment of the possible*. Individual genes do not produce traits. Traits are *outcomes* of interactions of which genes are essential components. It is not possible to inherit an outcome. If you are uncertain about the correctness of that last statement, just think about inheriting a recipe. Do you feel that possession of the recipe *assures* a particular result?

Was Drosophila *Worth It?*

Biologists are sometimes asked if all the effort which was expended on *Drosophila* was prudent. This is an extremely important question. It reveals the misconception that it was the insect which was the object of the study. It was not. What was involved was learning how genes are arranged on chromosomes, how they interacted with one another and the environment, and how they mutated. *Drosophila* was a convenient stand-in for all the organisms on earth, ourselves included. From the knowledge gained has come the power to examine any chromosome, locate any gene, and ultimately to determine the physical arrangement and organization of the continuitive mechanism that binds the generations together.

The efforts of researchers from Mendel to Morgan established that the gene existed; it had reality in space and time. It linked the generations and was at the core of the moment-to-moment control of life's homeostatic state. The search for the nature of life had narrowed and would remain sharply focused throughout the remainder of the 20th century. The chemical nature of the gene and the way this chemistry accomplished the production of biological structure and function became the target of those scientists who would ultimately come to be called molecular biologists.

CHAPTER ELEVEN

THE DISAPPEARANCE OF NATURE

*"One eye, one bulging eye, the technological scientific eye was willing to count
man as well as nature's creatures in terms of megadeaths.
Its objectivity had become so great as to endanger its master,
who was mining his own brains as ruthlessly as a seam of coal."*

— Loren Eiseley

W hen Gregor Mendel devised an explanation for inheritance based upon a hypothetical "factor," was he providing something essentially different from a "power" or a "force"? Since the time of the Greeks, hadn't scholars been resorting to unseen and unknowable mechanisms of all sorts? What is the difference between a supposed "gene" and a genie? Both are endowed with the ability to accomplish wonders.

There is, however, a quality to the Mendelian gene which sets it apart from most of the other explanatory forces. Even though the gene's chemical identity remained unknown, its *behavior* was remarkably predictable. Using the symbolic representation (**A** and **a**) and the statistical treatment appropriate to randomly assorting particles, geneticists had provided convincing evidence of the presence of this unseen piece of matter.

It is important to emphasize a subtle but critical distinction between the concept of the gene as influencing only such superficial qualities as the color of eyes while the eye itself continues to be thought of as the product of other forces. The genes, collectively, were now seen to be involved in the transmission of the basic plan of life — not only eye color but eyes, not only eyes but eye sockets, not only sockets but skulls. Genes were the mechanisms by which life's patterning flowed through the generations.

Unity and Diversity

When I was a graduate student, the teacher who most strongly influenced me once said something that shocked me. Following a discussion of various kinds of research and the attitudes of researchers, H. Burr Steinbach said, "The trouble with many biologists is that they love animals." I knew Burr Steinbach to be an extraordinarily fair-minded man, and this remark seemed totally out of keeping with his generous nature.

It was only after I thought about the context within which his statement had occurred that I realized the significance of what he had said. Mendel had studied peas and Morgan had studied *Drosophila* not because of any *intrinsic* interest these organisms had for them but rather as means to an end. It may very well be that Darwin truly enjoyed studying birds for their beauty and insects for their fascinating social behavior, but he ultimately came away from the study of the creatures with a single *unifying concept*, evolution by the process of natural selection.

What Burr Steinbach had in mind was the difference between studying living *things* as contrasted with studying *life itself*, its very essence, not its manifestation in any given creature at any given moment. I know that Burr

Steinbach deeply loved the natural world and appreciated its variability and complexity, but for most of the history of the study of the living state the examination of that diversity had obscured the unifying qualities. It was first the cell theory and, later, the physiological perceptions championed by Claude Bernard which began to center the attention of biologists upon those qualities which were common to all creatures. The behavior, in transmission patterns, of Mendel's particulate factors was a strong unifying force since the patterns held true in mice and men as well as flies and garden peas. The gene was the single strongest piece of evidence for the essential unity of life. If *life* was ever to be satisfactorily explained, the nature of the gene and its action seemed the critical place to start. It is in this sense that genetics played the central role in biological study during the middle years of the 20th century. Completing Burr Steinbach's thought, while it is quite understandable that people like animals, the gene is not a naturally lovable entity.

Maturity in Science Involves the Unifying Process

"What impresses us about a truly satisfying explanation is its capacity to bring a diverse set of observations into harmonious *unity."*

What impresses us about a truly satisfying explanation is its capacity to bring a diverse set of observations into *harmonious* unity. It is possible to strike any combination of several keys on a piano simultaneously, but only certain combinations produce *harmony*. We know it when we hear it. There is a oneness and a compatibility to harmony, a symmetry which has not been imposed but which evidences itself. Harmony in scientific explanation isn't as easily perceived because the intellect is less automatic in its discrimination than is the ear. But the informed intellect knows harmony when it encounters it. That is what this chapter is about.

As with any story which has many components, deciding where to begin and how many parts to include presents a problem. Begin at the beginning is usually sound advice, but, as you will see, the true beginning wasn't made clear until several later contributions were in place. And as to how many strands to include in the telling, that depends upon how heavy a burden the thread is asked to bear. We will try for a thread that is strong enough to make the explanation fully satisfying without becoming a deadening burden.

Bandages and Salmon Sperm

The surgical clinic in Tübingen, Germany was a convenient place from which to obtain discarded bandages soaked with pus, but why in the world would anyone want to? In 1868 a 24-year-old Swiss, Johann Friedrich Miescher, had gone to Tübingen to study with Europe's most prestigious medicinal biochemist, Ernst Felix Hoppe-Seyler. Along with the other biochemists of his day, Hoppe-Seyler classified the major organic chemical constituents of living tissue in three broad categories:lipids, carbohydrates, and proteins. Miescher had a fascination with the cell nucleus and was seeking a source of pure nuclei so as to conduct a chemical analysis of their contents. He hit upon the idea of collecting the fluid pus and removing from it the white blood cells which it contained in very large numbers. These cells had the advantage of possessing relatively large nuclei and very little cytoplasm.

The material Miescher extracted from the nuclei was a mixture of protein and what we now call nucleic acids. Miescher referred to the mixture as **nuclein**. The material was quite unlike any of the other three major organic constituents, and Hoppe-Seyler insisted upon personally repeating his student's work before permitting its publication. By 1869 Miescher was able to report on the chemical composition of this material, and he noted that in addition to the presence of nitrogen and sulfur (well known to be present in proteins), nuclein was remarkably rich in phosphorous. With the publication of Miescher's studies, a fourth class of organic substances joined the classic three — lipids, carbohydrates, proteins, and now nucleic acids.

In 1870, Miescher returned to Switzerland and established his own research program in Basel which is at the headwaters of the river Rhine. Just as the Tübingen clinic's bandages had provided a convenient source of pus cells, the Rhine's male salmon provided Miescher with pure suspensions of sperm cells. Sperm have notoriously large nuclei in proportion to their almost non-existent cytoplasmic content. Using this material to continue his investigations, Miescher isolated almost pure **nucleic acid** which was later given its name by one of Miescher's students, Richard Altmann, in 1889.

Strand One: DNA and Its Chemical Characterization

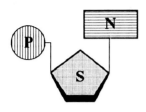

The three components of a nucleotide; phosphate, a 5-carbon sugar, and a nitrogenous base.

During the years between its isolation and the eventual clarification of its chemical structure in 1953, biochemists added to Miescher's original information concerning the nucleic acids. It turned out that there were two kinds. Both consisted of three components: a five-carbon sugar, a phosphate group, and what came to be called a nitrogenous base. One of the kinds of nucleic acid (the one discovered in the nucleus by Miescher) utilized a sugar called deoxyribose, and this component gave the compound its name, **deoxyribose nucleic acid (DNA)**. As the studies of DNA continued it became clear that the molecule was a very long one. It was a **polymer**. Molecules which consist of successive repeating units (like beads on a string) are called polymers. The individual units are called **monomers**. Each monomer in DNA consisted of one deoxyribose, one phosphate group, and one nitrogenous base. These three components formed a **nucleotide**. DNA consisted of a chain or string of successive nucleotides. The only variable quality seemed to be that four different nitrogenous bases existed: adenine, thymine, guanine, and cytosine. The reason for introducing this degree of chemical detail is not that every educated person should know the nature of DNA (although a case for that position certainly can be made) but because without this level of detail it isn't possible to follow what eventually happened. And what happened and why it happened *should* be understood by every educated person.

The second kind of nucleic acid utilized the five-carbon sugar ribose instead of deoxyribose. It came to be called **ribonucleic acid (RNA)**. It also was a polymer; each of its nucleotide monomers contained one ribose, a phosphate group, and a nitrogenous base. The bases were slightly different. Guanine, cytosine, and adenine were present but thymine was not. In RNA the fourth base is uracil.

From the very start, most of the investigative effort centered on DNA rather than RNA. There were two reasons for this. In general, science chooses to study substances for two quite different reasons. The first has to do with the perceived role or roles that the substance might play. The second is concerned with the ease with which a substance can be obtained and studied. Science, like all other human endeavor, must function in a real world. Scientists do what they can do. First, in the case of the nucleic acids, DNA was *theoretically* more "interesting." DNA was found in the nucleus, and the nucleus was assumed to be the location of the gene. RNA was found largely in the cytoplasm and was therefore more confusing in terms of any roles it might theoretically play. Second, RNA was more difficult to isolate and study.

The most influential person in the "chemical" part of this story was a biochemist named Phoebus Aaron Levene. P. A. Levene was a dominating figure in American biochemistry. His laboratory was located in New York at the Rockefeller Institute for Medical Research (now Rockefeller University). This institution was to provide the stage for many of the scenes in our story and P. A. Levene was to be a star performer.

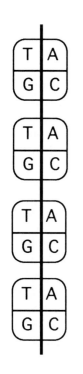

Levene's chemical analyses of DNA, obtained from a variety of organisms, showed roughly equal amounts of adenine (A), thymine (T), guanine (G), and cytosine (C). This information led him to propose a **tetranucleotide** structure for DNA. Tetra = four, and Levene's idea was that each nucleotide monomer consisted of deoxyribose, a phosphate group, and the *four* nitrogenous bases A,T,G, and C. Using our analogy of a string of beads, Levene's DNA molecule was a string of beads in which *each* bead was *identical* to all the others. Each bead consisted of the sugar deoxyribose, the phosphate group, and *all four of the bases.* This model explained, to Levene's satisfaction, the roughly equivalent amounts of A,T,G, and C that his analyses had revealed.

A tetranucleotide structure in which each unit is identical to all the others.

Strand Two: The Supremacy of Proteins

Biochemists tended to specialize in the objects of their investigations. When I was a graduate student in biochemistry, I took courses titled, "Proteins," "Lipids," "Carbohydrates," "Enzymes," and "Vitamins." Students are very sensitive to the pecking-order in their major departments — the relative prestige of various courses and professors. There was no doubt that in the biochemical world of the late 1940s proteins, and particularly the enzymatic proteins, were preeminent.

The reason for this was easy to see; proteins were perceived as being the primary functional as well as structural molecules. Take something like heart muscle. The contractile fibers which give to muscle cells their ability to shorten and lengthen are protein, and the enzymes which catalyze the energy flow in muscle cells are proteins. Biochemists had made remarkable progress in identifying the enzymes which specified the various energy-flow pathways which powered life and the details of exactly how muscle contracted pointed to the proteins as being the critical chemical molecules in life's mechanistic explanation. The key to the order and precision of this mechanism was the concept of **specificity**. We have considered this topic in some detail in Chapter Nine.

"In order to be specific a thing has to be identifiably unique."

In order to be *specific* a thing has to be *identifiably unique.* As you look at your keys you can identify your car keys from your house key by shape or color or some combination of factors. Each key has its own unique structure. This permits each of them to open only *its* complementary lock with no confusion of function.

Each enzyme depends upon this uniqueness of structure to catalyze only the reaction which enables its structure to interact with its specific substrate. The key analogy is additionally informative at this point. If we had only two locks to concern ourselves with it would not be difficult to design two differently shaped keys. Two short pieces of metal with alternative shapes will do. Let's increase the number to 100 locks, each requiring a specific key. Now the problem is more daunting. How many different ways can we cut the little up and down jiggles in a key so as to produce 100 unique versions? How about 1000 different keys? Intuitively we see that the keys will have to get longer so as to provide for more locations for the up/down alternatives.

Biochemists knew that they had isolated hundreds of different enzymes and that the cells had thousands more as yet unstudied. In every case the enzyme consisted of a protein portion and sometimes a smaller "coenzyme" portion which was frequently a vitamin derivative. Proteins were found to be remarkably suited to the role of providing specificity. Proteins, like nucleic acids, are polymers, chains of smaller monomer units. The protein monomers are **amino acids**. Each protein consisted of a long amino acid chain. And each chain was *unique* in its structure.

If we were told to prepare 100 strings of beads so that each string was identifiably unique, how might we go about it? First we'd ask if the lengths had to be identical. Suppose the answer was yes: all strings had to be 1000 beads long. Next we'd ask if all the beads available were identical. Now the answer is no; we have a supply of 20 differently colored beads. Are we free to use any color bead in any position? Yes, we can, and we can repeat the same color as many times as we wish. In fact, we can assemble the 20 differently colored beads in any patterns we desire. We'd smile and go about the assignment assured that we had ample means to make not only 100 different patterns but thousands, perhaps millions of different patterns. We'd ask a mathematician friend to calculate the number of ways 20 differently colored beads can be assembled in unique sequences 1000 units long. We'd know it was going to be a huge number.

etc.,
etc.,
etc.

Proteins consist of chains of amino acids, and there are 20 different amino acids typically encountered in proteins. The chains range in length from hundreds of amino acid units to thousands of amino acids. The variety of possible amino acid **sequences** is truly enormous. The *specificity* of enzymes was explained by the diversity provided by amino acid sequence differences. The molecular basis of specificity that biochemists discovered in protein structure was to dominate their thinking about the nature of the gene.

Strand Three: Death and Dying in 1923

Most of us have never known anyone who died as the result of pneumonia. In our minds, death comes from cancer, heart disease, AIDS, and drive-by shootings. The major killer in 1923 was pneumonia. Unless we know this, we cannot understand what was going on in the laboratory of Frederick Griffith at the Ministry of Health in London. Pneumonia was, for Griffith, cancer, heart disease, and AIDS all rolled into one.

It was known that the organism involved was a bacterium, *Streptococcus pneumoniae*, commonly referred to as pneumococcus. Griffith had learned that there were two major strains (variants) of this organism. One formed rough-appearing (R) colonies when grown in culture dishes, and under the microscope the individual bacteria appeared as simple spheres. This strain was quite harmless. The organisms were vulnerable to the defenses in animal hosts and could cause no disease. A second variant formed smooth-appearing (S) colonies in cultures, and under the microscope each individual sphere was seen to be enclosed within a gelatinous capsule which enabled the organism to avoid the defenses of a host. This strain was virulent; it killed.

Griffith was attempting to produce a vaccine and in the course of his efforts had injected a live virulent (S) strain of pneumococcus into mice. The injections were always lethal. He then heat-killed the S-strain organisms and injected them into mice. Now that they were *dead*, the S-strain organisms were harmless; the mice were totally unaffected by the presence of the dead S-strain bacteria in their bodies. Griffith also knew that the (R)-strain pneumococcus was not virulent and he could inject mice with *live* R-strain organisms and the mice would not be harmed.

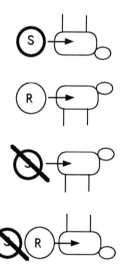

The results obtained following injection of live (S) strain, live (R) strain, dead (S) strain, and the combination of dead (S) and live (R) strain pneumococcus.

In what was to become one of the most famous experiments in biology, Griffith injected heat-killed, *dead* S organisms and *live* R organisms into the same animal. From what he had previously learned, that dead S organisms are harmless and live R organisms are harmless, the result of this double injection *should* have been "harmless." It wasn't. The mice died. When Griffith examined their tissues he found *live* S organisms in great numbers. There was no doubt about it; under the microscope there were the capsules, the mark of the S strain. And the organisms were alive. Injected into other mice they killed them off with undiminished virulence. And, most importantly for our genetic story, the offspring of these S-strain bacteria continued to be S-strain down the generations. The change was inherited.

Griffith had two alternative explanations at hand. Perhaps the dead virulent S-strain organisms had somehow been restored to life? This was not very likely in a mechanistic world. The other alternative was that the live, harmless R-strain organisms had somehow been **transformed** into the virulent S-strain pneumococcus! It is important that you remember that the S-strain organisms had offspring which remained S strain.

The transformation process wasn't at all well understood in the late 1920's. We now know what had occurred. When the S-strain cells were heat-killed, many were fragmented and pieces of their genetic substance were released. These pieces, groups of genes, were picked up by the live R-strain cells which incorporated the pieces into their own genetic apparatus. If an R cell received the genes which enabled the production of the gelatinous coat, the R cell began producing that coat. The R cell would now behave like an S cell; it would form smooth colonies. If injected into a mouse it could escape its defenses; it would multiply in the mouse and eventually kill it. The R cell was no longer an R cell. It had been transformed into an S cell. When the new S cell divided, it made copies of *all* its genes, the old ones and the newly acquired ones. Each daughter cell was an S cell.

The transformation wasn't superficial or temporary (like a brunette wearing a blonde wig); it was an inheritable *genetic* change which flowed down the generations. Griffith published the results of his studies in 1928, and the biological community was aware of what had occurred. For reasons we will discuss shortly, the community was not particularly impressed.

Are Bacteria to Be Trusted?

When news of the transformation of pneumococcus reached biologists, in general the results were viewed with suspicion. After all, it was as if one had put a cooked chicken and a live duck into a pen and out walked a live chicken which laid eggs out of which hatched other live chickens. Nothing less than the evident truth of **species constancy** was under attack. Was a sudden and inheritable transformation of one kind of creature into another possible?

The least troublesome way to deal with Griffith's results was to assume that bacteria possessed a primitive genetic system which did not operate with the stringency and precision exhibited by the higher organisms which followed Mendelian patterns. This permitted the evidence to be both accepted as scientifically valid but ignored as irrelevant in the general search for the nature of the gene. And that is exactly what happened in the thinking of the majority of researchers who used so-called "higher" organisms in their investigations.

What Is the Substance Responsible?

The majority is not everyone. Oswald T. Avery was a physician working at the Rockefeller Institute. He was aware of Griffith's work and also knew that essentially the same results had been reported shortly thereafter by researchers in Berlin. The implications disturbed him greatly because Avery was convinced that bacteria were indeed genetically like other organisms; in particular, he was positive that strains of bacteria did not undergo spontaneous conversions into different types. He repeated Griffith's work and, in 1929, confirmed the results. This set Avery on a path from which he would not turn for 14 years. Convinced that bacteria could indeed be genetically transformed, the constantly repeated question posed by Avery was, "What is the substance responsible?"

"What is the substance responsible?"

During the long search for the "transforming factor," a number of associates worked with Avery. One of the first, James Alloway, showed that transformation wasn't something that had to occur inside the body of a mouse. When heat-killed, virulent S-strain bacteria were mixed with live, harmless R-strain bacteria in a test tube, transformation occurred. This was a very important finding because it removed the veil of mystery from the transformation process. So long as transformation of the pneumococcus occurred inside a mouse, a complex living organism, the process remained hidden and its mechanism probably unknowable. By moving the transformation activities into a test tube, it became possible to follow all the events and study them.

Shortly thereafter, Alloway discovered another critical fact. He broke the virulent S cells open and passed the fragments through a very fine filter thus screening out pieces of the cell membrane and other large fragments. The fluid which passed through the filter was then tested to see if it could transform R cells into S cells. It could! The transforming factor was in the fluid! It could be isolated and identified!

This was easier said than done. The interior of a cell, any cell, is an incredibly complex mixture of chemical substances. Was it possible to identify which one of them was the transforming factor? Recall that biochemists classified the organic molecules into four classes: lipids, carbohydrates, proteins, and, following Miescher's discovery of them, nucleic acids. Lipids were readily extracted, and they proved not to be capable of accomplishing transformation. Carbohydrates were also eliminated as the potential genetic substance. The search narrowed down to the proteins and DNA. And here Avery and his co-workers Colin M. MacLeod and Maclyn McCarty ran into what researchers call "technical difficulties." The proteins and the nucleic acids proved to be impossible to completely separate from one another. What seemed to be a very simple and straightforward experiment became a laboratory nightmare.

"What seemed to be a very simple and straightforward experiment became a laboratory nightmare."

The idea was indeed straightforward. Extract the fluid interior of S cells, separate the protein from all the other substances, be certain you have absolutely pure protein, and expose R cells to the protein and determine its ability to transform them into S cells. Do the same thing with pure DNA. See which one of the two is capable of accomplishing transformation. The difficulty was that chemical efforts to completely separate the proteins from the nucleic acids were unsatisfactory. Mild chemical treatment seemed to leave traces of protein associated with the nucleic acid portion. More rigorous treatment simply destroyed the integrity of the molecules.

Civil War at the Rockefeller Institute

During the years when Avery and his associates were trying to learn the chemical identity of the transforming factor, it was clear that they favored the nucleic acid alternative. This was unacceptable to a number of other scientists at Rockefeller. The biochemist P. A. Levene remained active there until his death in 1940, and he insisted, on theoretical grounds, that DNA could not be the transforming factor. Levene's tetranucleotide structure for DNA portrayed the molecule as a string of identical monomers. Such a molecule couldn't possibly convey specific information. Keep in mind that each kind of genetic material must have its own specific structure. If the gene which specified the R form was identical with

"Levene's tetra-nucleotide structure lacked the ability to convey information. *It was like a code which simply said dot, dot, dot, dot, without any varia-tion."*

the gene which specified the S form, how could the two strains ever differ from one another? Levene and others speculated that DNA might be some sort of container or supporting structure for the gene but couldn't possibly be the gene itself. In current terminology, Levene's tetranucleotide structure lacked the ability to convey *information.* It was like a code which simply said dot, dot, dot, dot, without any variation.

Proteins, however, were ideally suited to contain infor-mation. We've already discussed the specificity of pro-teins: the fact that the 20 different kinds of amino acid monomers can be assembled in an almost infinite number of sequences. On theoretical grounds, proteins seemed the perfect candidate to be the substance of the gene.

At the Rockefeller Institute the foremost champion of protein as the genetic material was Alfred E. Mirsky. Mirsky was another eminent biochemist who argued against Avery's efforts to demonstrate that DNA was the genetic substance responsible for transformation. Mirsky took particular aim at the technique which Avery was attempting to use to prepare pure, uncontaminated DNA. Avery and his colleagues had finally decided that it was probably impossible to chemically *separate* the protein from DNA. The two substances seemed to be inextrica-bly combined. But there was an alternative to separation. Avery had decided to *destroy* the protein while leaving the DNA unchanged. He was counting on enzymatic specificity.

It was known that there were enzymes which catalyzed the digestion of various substances. In the stomachs and intestines of animals were found a variety of enzymes involved in the digestion of various foods, such as **pro-teases** in the digestion of proteins, and amylases in the digestion of starches. Avery's group set about treating the DNA/protein mixture with enzymes.

Starting with a protein/DNA mixture obtained from S cells and which clearly contained the transforming factor (in other words, was able to transform R cells into S cells), the scientists added protein-digesting enzymes. In theory, the material which escaped digestion should be DNA. Specificity would limit the action of a protease to the digestion of proteins.

Indeed, the remaining material *did* accomplish transformation! In an effort to be totally convincing, Avery used an enzyme, a so-called DNase, which catalyzed the digestion of DNA. When DNA was digested all transforming activity was destroyed. In addition, the possibility that RNA might be involved was explored by using an enzyme, RNase, to destroy any RNA which might be present. Avery repeated these tests for several years, using a variety of different enzymes and techniques. All the evidence pointed to DNA as the transforming factor.

Alfred Mirsky was unconvinced. Starting from the theoretical position that DNA simply could *not* be a genetic material since it lacked the specificity required for an informational molecule, Mirsky threw his considerable prestige behind an attack on Avery's work. The basic flaw, Mirsky insisted, was that the enzymes which were supposed to have destroyed *all* the protein in the DNA samples had failed to do so. Some protein had survived, insisted Mirsky, and it was this protein which accomplished the genetic transformation. Every time DNA seemed to have been identified as the genetic material, Mirsky would point to P. A. Levene's molecule and ask, "How can a non-specific, information-lacking molecule like DNA possibly accomplish a genetic transformation?"

"How can a non-specific, information-lacking molecule like DNA possibly accomplish a genetic transformation?"

In February 1944, Avery, McCarty, and MacLeod published the results of their many years of effort. "Induction of Transformation by a Desoxyribonucleic Acid Fraction Isolated from Pneumococcus Type III" is not the kind of title likely to attract an avid readership. To put things into a historical context, World War II was raging in Europe, Germany would not surrender until May 7, 1945 and the Japanese would continue to fight until August 14, 1945. Perhaps if the article had been titled, "Chemical Nature of the Gene Discovered!" it might have been noticed more widely, but Avery was a perfect example of the thorough, cautious, and self-effacing scientist. Regardless, the significance of the article was recognized by several researchers whose investigations were strongly influenced by it.

World War II and the Bacteria-Eaters

History is rarely given much of a role to play in the

teaching of science. It's true that books provide dates and teachers mention that Darwin's trip was on a sailing ship, but that isn't what I have in mind.

"Were it not for Adolph Hitler and the Nazi persecutions, it is very unlikely that a particular group of European scientists would have fled their homes and assembled in the United States."

Were it not for Adolph Hitler and the Nazi persecutions, it is very unlikely that a particular group of European scientists would have fled their homes and assembled in the United States. Max Delbrück was a German physicist who turned to biology as did a number of other physical scientists who contributed greatly to establishment of molecular biology. He arrived at the California Institute of Technology in 1937 and began to study a virus.

Most of us would assume that the virus Delbrück was studying caused some significant human disease. We are puzzled to learn that the virus, a so-called **bacteriophage** (phage = to eat), infects only bacteria. The shorter term **phage** is commonly used. Searching for meaning we might guess that the bacteria involved are quite harmful to humans and the virus was being used to control this bacterial scourge. We are wrong again; the bacteria preyed upon by these bacteriophage, *Escherichia coli* (*E. coli*), are normal inhabitants of the human intestine and are usually harmless to humans. We are stumped but only for a minute. Of course, we conclude, it is characteristic of a scientist to study some absolutely worthless topic while the world is going up in flames.

Max Delbrück, the physicist turned biologist, continued to think like a physicist. He asked the basic question, "What is life?" in the same way that his predecessors had asked, "What is matter?" To get answers to bedrock questions one digs down. The simplest possible being that exhibits any of life's basic characteristics is the best target for bedrock questioning. Viruses multiply; they are replicating beings. They can be assumed to possess genes. Let's find out if they do.

Together with Salvador Luria, an Italian physician turned biologist, Delbrück established what has become known as the "phage group." This loose assemblage of scientists, who all used bacteriophage viruses as their objects of study, established what you and I call **molecular biology**. Every single piece of information used in genetic engineering and the biotechnology industry traces back to the phage group and its founder, Max Delbrück.

One of the early discoveries was that bacteriophage consist of two basic components — an outer coat consisting of protein and an inner core consisting of DNA. Using this basic piece of information, two phage group scientists, Alfred Hershey and Martha Chase, would contribute a telling piece of information in the search for the chemical nature of the gene.

Knots in the Web

I have referred to scientific explanations as having a web-like quality in that various pieces fit and support one another when the explanations are basically sound. With Avery's announcement that genetic transformation in pneumococcus was accomplished by DNA, a flurry of DNA activity ensued. In Strasbourg, France, André Boivin and Roger and Colette Vendrely reported a fascinating result of their studies. According to Mendelian transmission theory, every cell in the body (with one important exception) would have two genes for each trait, one having been inherited from the male and one from the female parent. The cytological investigations of the late 19th and early 20th centuries had confirmed that in typical body cells chromosomes occurred in pairs. Such cells (called **somatic**) are said to be **diploid**. The important exception, of course, being that the gametes, eggs and sperm, would have only one copy of the gene. Indeed, in such **germ** cells only one member of each chromosome pair is found. Germ cells are said to be **haploid**. The act of fertilization, in which two haploid gametes fuse, restores the diploid number.

Boivin and the Vendrelys measured the amount of DNA in somatic and germ cells in a number of different species. They found that in every case the amount of DNA in germ cells was *half* the amount found in the somatic cells of that species. The fit was perfect; Mendelian particles, chromosomes, and DNA content followed exactly the same pattern. The web was strengthened with each new piece of evidence.

Hershey and Chase performed an experiment in 1952 which one would have thought would put an end to the debate as to the chemical identity of the gene. It was a masterpiece of experimental design. The outer coat of bacteriophage, as we have said, is made of protein. The inner core consists of DNA. When a bacteriophage virus infects an *E.coli* cell, which of these two regions contains the genetic information necessary to permit the replication of additional viruses? Hershey and Chase raised two populations of viruses in radioactive media. They labeled one group with ^{32}P, radioactive phosphorous. The other group was labeled with ^{35}S, radioactive sulfur. Here's where a little chemical knowledge helps. Proteins contain some sulfur; they contain absolutely no phosphorous. DNA contains lots of phosphorous (in the phosphate group) but absolutely no sulfur. Now the two chemical parts of the viruses were **labeled** with radioactive tracers.

The two virus groups were allowed to infect *E. coli* cells and Hershey and Chase examined the infected cells. Only the ^{32}P-labeled DNA was found inside the infected cells. The core of DNA had entered the *E. coli* and infected them. The ^{35}S-labeled protein coat material was found outside the cells; it never entered. The only substance which could possibly be replicated, the viral genes, was made of DNA.

There were some hold outs in the protein camp. Mirsky wondered if viruses were good objects for the study of genes; after all, viruses could only replicate inside the cells of another organism. But the tide had definitely turned in favor of DNA.

But there remained one unsettling piece of information. Levene's tetranucleotide model of the DNA molecule.

Erwin Chargaff, the Man Who Found the Key

In New York, working at the College of Physicians and Surgeons of Columbia University during the same period that Avery was in the same city, another fastidious biochemist, Erwin Chargaff, was examining cellular chemicals. Chargaff was studying lipoproteins, the combination of lipids and proteins which play decisive roles in many cellular activities. The news from Avery's lab, that DNA was the genetic substance, made Chargaff change direction. He had worked a little with DNA and knew that Levene's tetranucleotide structure for it was based largely upon the fact that roughly equal amounts of the four bases — adenine, thymine, guanine, and cytosine — could be found in the molecule. Chargaff went to work repeating some of Levene's investigations. Using tissues from a variety of animals, plants, and bacteria, Chargaff did a meticulous job of measuring the amounts of A, T, G, and C that their DNA contained. He found that Levene had been wrong.

Levene reported the amounts as being "roughly" equal. Chargaff found, for example, that human DNA contained the four bases in these proportions: 28 parts adenine, 19 parts guanine, 16 parts cytosine, and 28 parts thymine. These are not "roughly equivalent" amounts and certainly are not the kind of evidence which should lead one to propose a tetranucleotide structure.

"[T]he ratios...of adenine to thymine and of guanine to cytosine, were not far from 1."

Adenine and guanine are classed as purines while thymine and cytosine are pyrimidines. The two purines constituted 47 parts, and the two pyrimidines constituted 44 parts. Chargaff puzzled over his numbers and ended up noting that the total amount of purine was close to the total amount of pyrimidine. And his paper, "Chemical Specificity of Nucleic Acids and Mechanism of Their Enzymic Degredation," published in 1950, contained one remarkably important comment; "[T]he ratios...of adenine to thymine and of guanine to cytosine, were not far from 1."

"Chargaff had the key but he didn't know where the door was. Watson and Crick found themselves in front of a challenging door and recognized that Chargaff's key would open it."

As you and I read Chargaff's words we can be excused for not immediately appreciating their significance. It seems that Chargaff didn't either. This phrase was to become the key to the puzzle of DNA's structure. But it was not Erwin Chargaff who picked up the key and unlocked the golden door. In later years, after James Watson and Francis Crick were awarded the Nobel Prize for their establishment of the structure of DNA (the event which totally changed the study of biology and committed its scientists to the molecular age), Chargaff would insist that it was he who had provided the crucial evidence. In this judgment he was technically correct, but unfortunately for Erwin Chargaff the world recognizes those who *use* evidence imaginatively and creatively. Chargaff had the key but he didn't know where the door was. Watson and Crick found themselves in front of a challenging door and recognized that Chargaff's key would open it.

The Self-Replicating Molecule

Life has all sorts of fascinating attributes but at the core of any mechanistic explanation of the living state is the problem of reproduction. Ever since the 19th-century discovery that all organisms are essentially cellular and that all cells are descended from preexisting cells, it has been clear that life simply doesn't pop into being spontaneously from inert materials. Life is a continuum; each *individual* may have its own beginning and end, but life itself, *livingness*, flows through the generations without ceasing. The fundamental problem for a mechanistic basis of life is to provide a convincing explanation for this continuity.

In the search for the genetic substance, it was this test which the chemical candidate for genehood would have to face and pass. Was it conceivable that any molecule could convey such a momentous message down through the ages? Perhaps you can understand the concerns that the protein chemists had about the information-carrying capabilities of DNA. A molecule which was to encode life's directions would have to be quite a message carrier. And if DNA was truly the genetic material, *it* would have to be capable of **self-replication**. How else would one explain how I can keep my genes to live my life and also make copies so that my children can come into existence and live theirs? As stated in the saying about ultimate responsibility, "The buck stops here," DNA must not pass the buck to some "other" explanation.

There is something scary about closing in on an explanation. All the evidence from Miescher through Griffith to Avery and ultimately to the phage group pointed to DNA as the genetic material. What if the substance of the gene turned out to have no qualities which were useful in arriving at a mechanistic explanation? What if DNA carried no information, was incapable of replicating, yet *was the gene*? What would mechanistic biologists do then?

An Unlikely Pair

The Cavendish Laboratory at Cambridge University in England was a place devoted to X-ray crystallography. It was a physicist's place, a physical chemist's place, to be precise, a place which examined the structure of large, complex molecules. Sir Lawrence Bragg, its director, had originated X-ray crystallography as the procedure best capable of determining the structure of large molecules. That was what the Cavendish lab did.

Francis Crick was in his middle thirties, a physicist who had turned to biology, and he was pursuing his Ph.D. degree under the direction of Max Perutz, the senior crystallographer at the Cavendish. Perutz had discovered how to use X-ray diffraction to study the structure of truly enormous molecules such as the proteins. The process consisted of aiming X-rays at crystals of the material being studied. The X-rays bounced off (diffracted from) the surfaces of the molecular structure, and were recorded on film. The patterns which the X-rays formed gave evidence of the structure. The most critical item was the crystal of the material being studied; purity was essential. Even after the pure crystals had been obtained, interpretation of the X-ray patterns was an extremely tedious and time-consuming process. It was not work for the impetuous or the undisciplined. Francis Crick was slogging his way through a study of the hemoglobin molecule. Francis Crick was extraordinarily bright. He was also bored with his Ph.D. work.

James Watson was 11 years younger than Francis Crick. He started his college years as a 15-year-old at the University of Chicago. Lots of his classmates were very bright 15-year-olds. In a course in genetics, Watson was introduced to the research of Avery and became interested in the nature of the gene. But more influential upon him was a book by the physicist Erwin Schrödinger, *What Is Life* in which Schrödinger speculated upon the molecular nature of the gene. Although this 1944 book did not identify any particular substance as being the gene, it did suggest some of the characteristics such a chemical would have to have. Earlier I said that Erwin Chargaff was a man with a key who didn't know what door it might fit. Watson had a door. He started searching for the key.

After leaving the University of Chicago in 1947, Watson applied for admission to Graduate School at Harvard which turned him down. He was also rejected by the California Insitute of Technology. He had also applied to the University of Indiana at Bloomington. The University of Indiana accepted him. The senior Professor of Genetics at Indiana was Hermann Muller who had been trained by T. H. Morgan. Muller won the Nobel Prize, in 1947, for his discoveries of mutations caused by X-rays.

Although Muller had used *Drosophila* for most of his research, as far back as 1921 he had speculated about using viruses as genetic models. Watson had been attracted to Indiana by the presence of Hermann Muller. After he arrived, however, he was influenced most strongly by Salvador Luria who was also at Bloomington. You may recall that Luria was one of Delbrück's phage group. It was Delbrück and Luria who were primarily responsible for introducing the combination of *E. coli* and its parasite, the bacteriophage, to the biological sciences. These two organisms were to become the basis upon which molecular biology was established.

At Indiana, Watson worked in Luria's laboratory and, through Luria, met Max Delbrück. During the summer of 1948 Watson participated, for the first time, in an experience which communicates a great deal about the way science was actually done in the 1940s and 1950s. Watson accompanied Luria and his wife to Cold Spring Harbor, Long Island, one of several east coast biological research centers which combined all of the best aspects of informal scholarship with a summer vacation. Unlike the highly structured nature of most universities, these summer stations stressed informal and personal relationships between senior investigators and their students. Cold Spring Harbor also hosted an annual international symposium which drew prominent scientists from all over the world. These were opportunities not only to exchange the latest in research findings but to establish working and personal relationships which were to last a lifetime. Through Luria and Delbrück, Watson was drawn into the center of the rapidly expanding international world of molecular biology. In the summer of 1951 Watson arrived at Cambridge. He immediately began a collaboration with Francis Crick. The race to the double helix was on.

Although James Watson's book, *The Double Helix* is a wonderfully readable, candid account of the discovery of the structure of DNA, a much better source of information is a remarkable book by Horace Freeland Judson, *The Eighth Day of Creation*. In his book, Judson reports on conversations not only with Watson and Crick, but with every significant researcher of the period. As a result, we are provided with the viewpoints of friends and associates as well as rivals and enemies. Science comes alive as the very human endeavor that it is.

One of the most fascinating conversations reported by Judson is one he held with Linus Carl Pauling who was, in 1951, the favorite in any betting as to who would discover the structure of DNA. After all, Linus Pauling literally wrote the book on the structure of large molecules. His text *The Nature of the Chemical Bond* was used in every research laboratory where young molecular investigators learned their skills. Francis Crick had learned his skills from Pauling's works. Pauling would eventually win two Nobel Prizes. The race to the discovery of the structure of DNA would be run on Linus Pauling's track, and Pauling was very confident of the outcome.

Pauling was not the only other investigator chasing the elusive prize. At King's College in London was Maurice Wilkins, another physicist who had become fascinated with biological molecules. Wilkins was also doing X-ray diffraction studies. Rosalind Franklin, a very capable crystallographer was hired to participate in Wilkins's studies. From the very beginning, Rosalind Franklin proved to be a very independent woman who was not in the least interested in cooperative efforts. She refused to consider herself as working with or for Maurice Wilkins. She conducted her own investigations and shared her results very reluctantly. Her results were critical to the outcome of the story.

"DNA, you know, is Midas's gold. Everybody who touches it goes mad."

Many years after the structure of DNA had been established and the bitterness and disappointment had lessened, Maurice Wilkins told Horace Judson about the atmosphere of the race. "DNA, you know, is Midas's gold. Everybody who touches it goes mad."

The Ways of Working

Working with proteins, Linus Pauling had established the fact that long polymeric molecules tended to coil; they formed what the physical chemists called a helix.

The coiled telephone cord is a pretty good approximation. The reason for the coiling is that the amino acids which are the monomers of a protein all possess chemical groups which form **hydrogen bonds**. In Chapter Eight we considered the role of such bonds in bringing about orderly structure, as in water. In polymers such as proteins, hydrogen bonds form between amino acids about four units apart in the protein chain. The attraction of these regularly spaced amino acids for one another forces the chain to assume a helical configuration. Pauling was convinced that DNA would also be a coiled molecule because its monomers, the nucleotides, also had the capacity to form hydrogen bonds with one another. It was the task of the X-ray crystallographers to determine exactly where all of the components of nucleotides were positioned — the phosphate group, the deoxyribose, and the nitrogenous base. Pauling was convinced that he could derive the structure from a relatively small number of pieces of information; it was a question of putting the right pieces together in the correct pattern. He depended upon his extensive knowledge of protein structure to carry him into the structure of DNA. Pauling was not only confident, he was disdainful of his competition.

Rosalind Franklin's X-ray diffraction pictures were the best produced by any of the various laboratories. Her instincts told her to keep perfecting the techniques she was using and get progressively better data. She was convinced that painstaking effort and getting increasingly revealing patterns would lead to the structure. She was emotionally opposed to speculations which were not based upon extensive and precise information. She worked with a dispassionate and disciplined precision.

This was not true of Watson and Crick. These two were of an entirely different temperament — bold, quick to see possibilities, and impatiently eager to gamble on their intuition. And they lived in constant fear that Linus Pauling would arrive at the solution while they were still thrashing about. Watson and Crick wanted very badly to win.

The Data and Its Use

Once a scientist has begun a project, the results of the investigation begin appearing. Like all results, some are tentative and will very likely require revision. Others are simply worthless — not necessarily wrong, but irrelevant. Not everything one learns is meaningful. Some pieces of information are critical to the solution of a problem. In the process of selecting from among the bits and pieces those which are essential to the task at hand, a scientist reveals what we tend to call creativity. Some scientists, like some artists, are more creative than others. Remember Werner Heisenberg's aphorism, "What we observe is not Nature itself, but Nature exposed to our method of questioning." Creative use of information, creative phrasing of questions to be asked, creative assemblage of Nature's responses, are what distinguishes intellectually significant science from routine data gathering.

What was publicly known about the chemical nature of DNA in 1951, the year Watson arrived at the Cavendish and Rosalind Franklin began work at King's College, is what you know from reading this book. DNA was a polymer; its monomers were called nucleotides; and each nucleotide consisted of the sugar deoxyribose, a phosphate group, and one of four nitrogenous bases. The bases were adenine, cytosine, guanine, and thymine. In addition, Chargaff's revision of Levene's tetranucleotide model had been published in 1950. What was not known, and what the race was all about, was how these various components were assembled.

Notice that I have said "publicly" known. There are two ways that scientists communicate the results of their efforts: publically, in printed reports in scientific journals or oral presentations at scientific meetings and, privately, through conversation and letters. It is this latter route that is often more important. By the time a discovery has actually appeared in print, it may have been privately revealed to a small group of close associates for many months. In a race it is critical not only to know but to know as soon as possible.

Rosalind Franklin's X-ray crystallography had progressed to the point where she felt pressured to present what to her were still inconclusive findings to a small group at King's College. Watson, alerted by Maurice Wilkins, attended the meeting. When he reported to Francis Crick the diffraction patterns he had seen, Crick was more than ever convinced that DNA was a coiled molecule, a helix. Furthermore, he thought it was probably a double helix, two strands coiled about one another.

Watson and Crick embarked upon a model-building effort. Making diagrams, drawn to scale, of the various known chemical consituents of DNA, they tried to assemble the pieces so as to produce a double helix. Their first model was a miserable failure. It violated all of the chemical bonding rules; the structure thay proposed was chemically impossible. It would literally fly apart.

They suspended their model building temporarily awaiting additional information. In an incredible turn of events, Linus Pauling's son Peter turned up at the Cavendish. He told Watson and Crick that his father was about to publish a paper announcing the structure of DNA. Watson and Crick were plunged into despair. They asked Peter Pauling to write to his father requesting a copy of the manuscript. He did so and Pauling sent him one. Peter showed it to Watson and Crick. Pauling's proposed structure was similar to an earlier one Watson and Crick had rejected as being impossibly wrong. Had they thrown away the prize? No, Linus Pauling had simply made some fundamental errors in basic chemistry.

Pauling had relied upon diffraction data produced in 1947, data which were crude and incomplete. From it, he had tried to construct a model of DNA which Watson and Crick knew was wrong in the light of the much more complete and accurate data which Rosalind Franklin had produced. The race was far from lost. The model building began again with a vengeance.

"The race was far from lost. The model building began again with a vengeance."

From the X-ray diffraction patterns, Watson and Crick knew the internal diameter of the helix. It was 2 nanometers from one strand to the other. A nanometer is one-billionth of a meter. Imagine a ladder which is known to be one foot wide between the two upright poles. Something must be holding the two uprights, in position, a foot apart. By using their scale models of the various DNA components, they constructed various versions attempting to place them into positions which would arrive at the 2 nanometer distance. The most promising model had the deoxyribose and phosphate groups forming the uprights of the ladder. That meant that the nitrogenous bases had

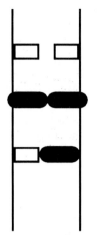

Two purines (solid ovals) are too long while two pyrimidines (open rectangles) are too short. One purine and one pyrimidine fit within the 2 nanometer distance between the uprights.

to fit inside, between the uprights, playing the role of the ladder's rungs.

Here it essential to recall that the four bases are of two sorts: adenine and guanine are purines while cytosine and thymine are pyrimidines. The modeling efforts showed that none of the four bases was large enough to play the role of a rung; none would stretch the 2 nanometer distance. This led to placing two bases between the uprights. If the two purines (adenine and guanine) were placed side by side, they were too long. If the two pyrimidines were placed side by side they were too short. But if one purine and one pyrimidine were placed side by side between the uprights, they spanned the 2 nanometer distance precisely. The excitement was building to a fever pitch.

Chargaff's Key to the Golden Door

It is time to recall Chargaff's comment about the bases, "[T]he ratios...of adenine to thymine and of guanine to cytosine, were not far from 1." Adenine is a purine and thymine is a pyrimidine. Guanine is a purine and cytosine is a pyrimidine. It requires one purine and one pyrimidine to stretch exactly 2 nanometers from one deoxyribose/ phosphate strand to the other. Chargaff's data revealed that there were equal amounts of adenine and thymine (that's what the ratio being "not far from 1" means). Similarly, guanine and cytosine were present in equal amounts. The best way to account for Chargaff's figures was to assume that if one-half of a rung was made of adenine, the other half must be thymine. If guanine comprised one-half of a rung then the other half must be cytosine.

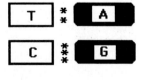

Thymine pairs with (complements) adenine, and cytosine pairs with guanine.

But what forces existed to hold the two halves of a rung to one another? In manipulating their model pieces, Watson and Crick found that between adenine and thymine there were two locations which would form hydrogen bonds. The bonds would form in exactly the right position to link the bases together. Between guanine and cytosine were three hydrogen bond-forming locations which gave a precise fit. The model was complete.

It consisted of two backbone strands made up of alternating deoxyribose and phosphate groups. These two strands

Following the separation of the original double-stranded molecule, each single strand serves as the guide for the synthesis of its complementary strand.

spiraled about one another to form the outer portion of the molecule. The interior consisted of the bases which bridged the 2 nanometer gap between two strands. The hydrogen bonds between the thymine-adenine pairs and the guanine-cytosine pairs held the double helix together.

The Implications of the Model

Nonscientists are sometimes confused when they hear a scientist speak of a scientific experiment as being "elegant" or when they learn that a theory is thought of as being "beautiful." These are not dispassionate terms. They imply an esthetic quality which sounds foreign to the supposedly unemotional scientific attitude. Perhaps my description of the race to the double helix wasn't sufficiently dramatic to convey how intensely the participants were competing not only with one another but with the unknown. Science is *not* dispassionate. Creative people are intensely committed to their creations. Sometimes a creation brings together all the elements of a problem and joins them in a solution which is so harmonious, so obviously correct, that it provides for the beholder the same sense of awe that one experiences upon first seeing the Rockies.

The Watson-Crick model evoked this kind of response from its first audiences. The others in the race, upon seeing it, knew instantly that it was correct; this *was* DNA. The model revealed the molecular basis of life's central mystery; reproduction. It was almost self-evident that if the hydrogen bonds between thymine and adenine and between cytosine and guanine were to break, the two strands of the DNA molecule would separate from one another. Each strand could then be the pattern for the creation of a new **complementary** strand. Wherever the original strand had a thymine, the newly forming strand would insert an adenine. Wherever the parent strand had an adenine, the newly forming strand would insert a thymine. The same would hold for guanine and cytosine. Only thymine-adenine and guanine-cytosine combinations could form the proper hydrogen bonds. DNA was capable of self-replication.

As to how the molecule could convey information, that also became clear after a little consideration. Recall that proteins are specific due to the *sequences* of their amino acids. DNA also had sequence specificity. The T,A,G,C sequence along the strands turned out to be a form of code. The bases can appear in any and all combinations; T-A-A-G-G-C-T-A-T, for example, on one strand would find A-T-T-C-C-G-A-T-A on the other strand. We now know that only one of the two strands is "read," and in Chapter 12 we will examine how scientists learned to read the code and how the code is expressed.

These two features, self-replication and information content, lie at the heart of the mechanistic explanation of life. Watson and Crick's model revealed the chemical basis of these critical features of the living state. With their model completed, biologists and chemists had arrived at the tower in the center of the enchanted forest. A double helix stairway wound up to the top from which the entire forest could be viewed as a unified whole.

But the forest had changed. It was no longer the incomprehensible tangle of diverse appearances of life. That view had disappeared and in its place was the crystalline unity of molecular design. The unifying pattern had been exposed by the scattering of X-rays and the ingenuity of imaginative speculation. From the vantage point of the double helix, every living thing could be interpreted as a variant of a primordial DNA sequence. The questions to be asked of Nature all centered on DNA, its structure, its method of replication, its message, and its evolution. The visible living world was seen to be the expression of the program encoded in the double helix.

Rewards and Challenges

In 1962 the Nobel Prize was awarded to Watson, Crick, and Maurice Wilkins. Rosalind Franklin was not included. In order to put this event in perspective, it is essential to know two important things about the Nobel Prize selection system.

The structure of DNA was announced in 1953. Why did it take nine years to acknowledge the discovery? The Nobel Committee considers a very large number of accomplishments and many significant ones had been awaiting their turns for recognition. The slow and ponderous process may seem unsatisfactory, but history has demonstrated again and again that many findings which seem momentous at the time fade into relative obscurity when additional research reveals more significant aspects. Some highly acclaimed scientific work has turned out to be in error. The slow pace of the Committee's deliberations permits a degree of perspective which a more hasty process would not.

But the absence of Rosalind Franklin's name seems unforgivable. It was Maurice Wilkins' laboratory at King's College at which the critical X-ray diffraction studies were performed, but it was Rosalind Franklin who did the decisive work. She left King's College in 1953 to work in the X-ray crystallography laboratory of John Bernal at Birkbeck College in London. She died of cancer in April 1958 at the age of 37. The Nobel Prize is never awarded posthumously; only living persons may receive it. The slow pace of the Nobel process resulted in a distortion of history.

"She died of cancer in April 1958 at the age of 37. The Nobel Prize is never awarded posthumously; only living persons may receive it. The slow pace of the Nobel process resulted in a distortion of history."

In speaking of his highly productive relationship with Watson, Francis Crick said that collaboration is essential in science. Well-matched partners prevent one another from falling into error, from pursuing false trails too long; a good collaborator, said Crick, stops the nonsense. In hindsight, it was the absence of such a collaborator which led to Rosalind Franklin's failure to see the path to which her data so strongly pointed. Her preserved notebooks reveal a continual uncertainty, a vacillation which the right partner might have countered.

Molecular biology blossomed after 1953. After two centuries of debate as to whether life could adequately be explained mechanistically, the structure of DNA suggested not only a strongly positive answer but indicated the wording of some of the most meaningful questions to be asked of Nature. As is usual with meaningful responses, Nature's answers raised questions not even dreamed of when the search for the gene began.

CHAPTER TWELVE

THE PRIDE OF THE GARDEN

"The joining of the two mainstreams of molecular biology and the fusion of both with biochemistry constitutes the 'molecular revolution' in Biology. So successfully have molecular biologists fulfilled their function that the need — except for pedagogic purposes — for a separate and distinct discipline of molecular biology may now be questioned. It is no longer a delicate plant that needs a special soil and anxious attention: it is now the pride of the garden."

— Peter B. Medewar

T here is a new biology. Medewar describes it as the result of the fusion of three streams — two flowing from molecular biology and the third from biochemistry. It will be helpful to sort out these streams before they become so intermixed in the "new biology," the pride of the garden, that we cannot see clearly where they came from and how they intermingled.

Starting with its discoverer, Friedrich Miescher, DNA had been the primary target of genetic interest, but its close relative, RNA, had always held a fascination for those scientists who studied the events occurring in the cytoplasm. While DNA's location in the chromosome gave it an obvious potential role in inheritance, RNA's primarily cytoplasmic location did not indicate any easily identifiable function. With time, however, one possibility arose: RNA seemed to be found in locations and amounts which varied with protein production. In cells which secreted very large amounts of protein (insulin-producing pancreatic cells, for example) the amount of RNA was similarly very great. Even more indicative of its relationship with protein production was the fact that RNA varied both in amount and location in a manner which directly mirrored protein formation.

"So here were the three tributaries, DNA, which had to replicate and contain information, RNA, which seemed somehow to be concerned with protein production, and protein itself, the molecule out of which enzymes were made. "

Proteins were the functional and structural molecules which biochemists had taken to their hearts. Since all known enzymes were protein in constitution, and since enzymes controlled the flow of metabolic commerce in and between cells, there could be no doubt that life's mechanism would be heavily dependent upon the highly specific proteins.

So here were the three tributaries, DNA, which had to replicate and contain information, RNA, which seemed somehow to be concerned with protein production, and protein itself, the molecule out of which enzymes were made. Investigators had learned quite a bit about each of them but as yet there was no coherence to the knowledge; it just didn't form a single stream. How were these substances linked to one another?

One Gene: One Enzyme

I remember a late-spring day spent trout fishing when I was walking along the edge of a large snow-field at about

10,000 feet in the Rockies. Above me, stretching up to the 12,000 foot peaks was an unbroken mantle of white. The melting snow at the edge of the field was forming rivulets which streamed off down the slope toward a small valley where a series of beaver dams temporarily slowed the run-off which I knew eventually would flow into Big Muddy Creek. The Big Muddy is a somewhat pretentious name for a rather unimpressive stream which wanders through grazing land and eventually flows into the Colorado River. As I watched a trickle of water fill my footprint, I thought of the momentary interruption I had caused to the torrent which would eventually flow through the Grand Canyon. It is very difficult to believe that it is the untold numbers of tiny trickles which are the ultimate sources of the river which has gouged that mile-deep gash in the surface of this planet.

The contributions to our understanding of the molecular basis of life are also frequently difficult to recognize as significant. In 1941, George Beadle and E. L. Tatum performed an experiment using the pink bread mold, *Neurospora crassa*. By this time we should be aware that investigators select biological models, not because they are necessarily interested in preventing moldy bread, but because the organism in question has some particularly useful characteristic. In this case, Beadle and Tatum selected an organism which has an extremely well-defined metabolism. *Neurospora* can grow on a medium (a so-called **minimal** medium) which contains only an energy source (a sugar), one vitamin, and inorganic salts. The organism can synthesize all of its own complex organic chemicals from the minimal diet. It can do so because its genetic constitution (its **genome**) has the capacity to form all of the enzymes necessary to execute the chemical reactions required in the synthesis of the organic chemicals.

Chapter Nine informed us that biochemists had learned that metabolic reactions occur in pathways, sequences of reactions in which a starting substance is modified, in a series of steps, into a final end-product. The requirement for specific enzymes permits each step to proceed only if its particular enzyme is present. Beadle and Tatum reasoned that a normal *Neurospora* must produce every enzyme required for the steps in the metabolic pathways which produce its end-products. It had been assumed that one of the crucial roles of genes was the production of enzymes but this relationship was just that, a conjecture. Nobody had ever demonstrated that genes did indeed produce enzymes.

"Beadle and Tatum speculated that if genes did produce enzymes it might be possible to demonstrate that the loss *of a gene resulted in the* absence *of an enzyme."*

Beadle and Tatum speculated that if genes did produce enzymes it might be possible to demonstrate that the *loss* of a gene resulted in the *absence* of an enzyme. They intended to use X-rays to create mutations in *Neurospora*, hoping that some of the mutations they caused would affect the genes which produced the metabolic pathway enzymes. The use of X-rays is like shooting a gun into a dark room; you cannot select your target. All hits and misses are purely the result of random chance. Beadle and Tatum exposed *Neurospora* to X-rays and did succeed in producing strains which could not grow on the minimal medium. The mutants lacked the ability to synthesize one (or more) of the important organic molecules.

In diagram form, the reasoning is as follows. Assume a metabolic pathway A $—>^1$ B $—>^2$ C $—>^3$ D $—>^4$ E. The substance "E" represents the end-product of the pathway. Each metabolic reaction is catalyzed by a specific enzyme indicated by the numbered arrows. If enzyme 1 is produced by gene 1, enzyme 2 by gene 2, and so on, then a mutation in gene 3 would produce a defect in the pathway: A $—>^1$ B $—>^2$ C $\sim>$. In the absence of a properly functioning gene 3 there would be no enzyme 3 and there would be no ability to produce substance D, thus bringing the pathway to a stop before the end-product had been produced. But it isn't possible to *see* metabolic pathways, and it isn't possible to *see* which gene has mutated. Recall that all mutations caused by the X-rays were random events.

Beadle and Tatum thought of a way to detect which enzyme was missing as the result of a mutation. In the example above, the reaction has stopped because gene 3 has mutated and is no longer capable of producing enzyme 3. The organism cannot survive on minimal medium. But suppose we were to add to the minimal medium just one substance, **compound D**! Even though the mutant, by itself, cannot make compound D, we've supplied it! Now the pathway looks like this: A $—>^1$ B $—>^2$ C $\sim>$ **D** $—>^4$ E. Compound **D** has been supplied in the diet, and since the only gene which has mutated is the one required for the production of enzyme 3, all the other enzymes, including enzyme 4, are present and will function. The *pathway* is restored even though enzyme 3 is missing. This mutant can survive on the minimal medium *supplemented* by the addition of D. Even though mutated, it can produce the end-product and can survive, but only on the enriched medium.

In extensive tests, Beadle and Tatum demonstrated that the X-irradiation caused mutational changes in the genes of *Neurospora*. Each mutant lost the ability to produce a specific enzyme. Apparently the role of each gene was to produce a specific protein product, like an enzyme. "One gene, one enzyme" became a rallying cry for those searching for the way in which genes functioned

"Apparently the role of each gene was to produce a specific protein product, like an enzyme. 'One gene, one enzyme' became a rallying cry for those searching for the way in which genes functioned."

DNA's Challenges: I. Replication

The 1953 Watson/Crick structure for DNA "suggested" the method of its replication. The two strands of the helix were held to one another by the hydrogen bonds between their bases: A bound to T, and G bound to C throughout the length of the molecule. If the hydrogen bonds released, the two original strands could separate and each "old" strand would provide the guide for the formation of its "new" strand. Each of the new strands would be **complementary** to its guide; that is, wherever the old strand exposed an A, the new strand would hydrogen-bond to it a T, and wherever the old strand exposed a G, the new strand would hydrogen bond to it a C. Chargaff's Rule — A with T and G with C — would assure that when the replication process was complete, *two* double helices of DNA would exist and both of them would be exact copies of the original double helix.

```
A – T
C – G
T – A
G – C
G – C
A – T
```

```
A • T        A • T
C • G        C • G
T • A        T • A
G • C        G • C
G • C        G • C
A • T        A • T
```

The sequence of the original strands (solid letters) specifies the sequence of the newly synthesized complementary strands (open letters).

In 1958, Matthew Meselson and Franklin Stahl of the California Institute of Technology showed that the "suggested" process actually occurred. Using highly technical procedures, they labeled with isotopes the materials out of which the new strands would be formed. Sure enough, each replicated double helix consisted of one "old" strand and its labeled, complementary "new" one. DNA was not only theoretically capable of serving as its own guide for replication, it actually did so. The molecular basis of reproduction was established.

In the years following 1958 we have learned that there are a number of enzymes involved in the process of replication. Some of these unwind the double helix, exposing the bases; others are involved in attaching the complementary bases and checking for accuracy, cutting out any mismatches. This last-named task is enormously important. It turns out that there are circumstances when A does *not* bind T and G does *not* bind C. There is a rare form of cytosine which *can* bind with adenine, for example. This error, if permitted to stand, changes the original sequence of the bases. As we shall see, such a change sets the stage for confusion in the information content of DNA. There is an enzyme which detects mismatched bases and eliminates them. The enzyme isn't intelligent or purposeful, it is simply specifically shaped for that particular circumstance. It cuts out mismatches in the same mindless manner that other enzymes digest proteins. One of the things we will have to constantly keep in mind is that we are talking about *molecules*. Let's not give to them intelligence or intention which they simply do not possess.

DNA's Challenges: II. Information Content

The fact that there were four different bases used in DNA's nucleotides was a case of "good news/bad news." The good news was that four different bases (which could be arranged in any possible sequence) provided more than ample material for any message that DNA might be asked to convey. After all, the Morse Code consists of only two elements, the dash and the dot. Using these in a variety of combinations and sequences, the entire content of the *Encyclopedia Britannica* can be communicated. DNA had four elements with which to work.

The bad news was that unlike the code which Mr. Morse invented and obviously knew how to translate, DNA's code was unknown and no code book was hidden waiting for some spy to discover. There was, however, a pretty strong hint as to what the code was intended to accomplish.

Proteins derive their specificity from the sequence in which the 20 different amino-acids appear down the

*"If we consider proteins to be strings of amino acid beads, somehow or other DNA had to **specify** the sequence in which the 20 different amino acid beads were strung."*

length of the chain. From the work of Beadle and Tatum it was known that the role of the gene was somehow to bring into existence the protein enzymes. If we consider proteins to be strings of amino acid beads, somehow or other DNA had to **specify** the sequence in which the 20 different amino acid beads were strung.

The simplest conceivable code would be one in which each individual nucleotide of DNA specified one amino acid. The problem with this level of simplicity is that there are only four DNA nucleotides — A,T,G,C — while there are 20 different amino acids. There were five times as many amino acids as there were nucleotides. Perhaps it took *two* nucleotides to specify an amino acid; AA might stand for valine, AG might be cysteine, GA might be argenine, and so forth. If one puts the four nucleotides into groups of two it turns out that no matter how one assembles them there are only 16 (4×4) combinations. Still not enough.

The next possibility is that it takes *three* nucleotides to specify an amino acid. We find that the number of combinations of three nucleotides is 64 (4×4×4). But we only need 20 since there are only 20 amino acids which appear in proteins. The coding idea seemed to be going to pieces.

It was Francis Crick who suggested that the code might be **degenerate**. Degeneracy, a physicist's concept, means that several different conditions amount to essentially the same thing. Perhaps each amino acid could be specified by more than one sequence of three nucleotides.

Several laboratories embarked on the effort to discover the code. Marshall Nirenberg and Heinrich Matthaei of the National Institutes of Health made the first breakthrough. In 1961, using a newly developed method, they synthesized an *artificial* **polynucleotide**, that is, a polymerized strand consisting of many nucleotides. Nirenberg and Matthaei started out with a pure population of the RNA nucleotide containing the base uracil. Perhaps you recall that in RNA this base (U) substitutes for thymine. It is entirely reasonable to ask why these workers used an **RNA** nucleotide rather than a **DNA** nucleotide. Remember the conviction that RNA was involved in protein synthesis? By 1961 it had become clear that this was indeed so. We'll explore the RNA-protein connection shortly, but let's get back to the coding problem.

Nirenberg and Matthaei prepared poly U strands; that is, strands of RNA in which the nucleotide containing uracil was the only monomer present. Repeated again and again, the strand was U-U-U-U-U-U-U throughout its entire length. No matter where the code started or ended, if it consisted of three "letters," it would have to be read as U-U-U, U-U-U, U-U-U. This artificial piece of RNA was placed in a mixture of cell components which had been shown to be capable of synthesizing individual amino acids into a polymer which, although shorter than the average protein, was a very lengthy molecule indeed. Such amino acid chains are called **polypeptides**.

The mixture of cell components was supplied with all 20 amino acids. The code was fed in, the synthetic apparatus was in place. What product would emerge?

The cell components made a polypeptide consisting of only one amino acid, phenylalanine (phe) repeated again and again and again. Feed in a nucleotide chain consisting of nothing but U-U-U-U-U-U and out comes a polypeptide chain phe-phe-phe. The first code word was known: the nucleotide sequence U-U-U specified the amino acid phenylalanine. The groups of three nucleotides were called **triplets**. You should be able to guess the next triplets to be examined. It was easy to make nucleotide strands consisting of the triplets A-A-A, C-C-C, and G-G-G. It turned out that A-A-A specified the amino acid lysine (lys); C-C-C specified the amino acid proline (pro); and G-G-G specified the amino acid glycine (gly). The easy part was over. How would one make *combinations* of nucleotides, triplets such as G-U-A or U-C-G? Making a predetermined nucleotide sequence isn't like picking up a desired bead with a tweezers and stringing it into place.

Over the next five years a number of laboratories around the world developed ingenious methods of creating synthetic RNA nucleotides using A,U,C, and G in well-defined sequences. Eventually, all of the 64 triplets were tested, and the amino acids which they stipulated were revealed. By 1966 the entire genetic code had been learned.

The Genetic Code Is Degenerate and Universal

Francis Crick had been right when he suggested that the code might be degenerate. It turned out that the amino acid phenylalanine was specified not only by the triplet UUU but also by UUC. Some amino acids are specified by as many as four different triplets. Leucine (leu), for example, is specified by CUU, CUC, CUA, and CUG. There is one amino acid which is specified by only one triplet: methionine (met) is coded for only by the triplet AUG. Keep this fact in mind. It will be useful when we consider exactly how the code is expressed.

The triplets of RNA nucleotides (**ribonucleotides**) were given the name **codon**. Of the 64 codons, 61 specified amino acids. Three of the codons did not specify any amino acid at all. UAA, UAG, and UGA seemed to have no "meaning." As we shall see, these three codons, while they do not specify any amino acid, are particularly important. In the grammar of life, these play the role of punctuation marks. As you read this paragraph you are aware of the letters; they contain "meaning." What you may not have consciously realized is that only by bringing each word to a close and leaving a space between the end of one and the beginning of another can the "meaning" of the letters be realized. UAA, UAG, and UGA are so-called "stop" codons. Wherever they appear in a sequence of codons, the cellular machinery **terminates** the message.

" 'If it's true of E. coli *is it true of me?' It turned out that the code is universal. Humans use the same codons as* E. coli *to specify the amino acids."*

The deciphering of the genetic code was accomplished using nucleic acids obtained from the organisms with which molecular biologists were most familiar. *E. coli* was the ideal choice since most of the knowledge concerning the enzymes involved in the synthesis of polypeptides had come from this bacterium. The question which loomed large in the minds of the scientists who worked on the code was, "If it's true of *E. coli* is it true of me?" It turned out that the code is universal. Humans use the same codons as *E. coli* to specify the amino acids. So does *Drosophila*. Pea plants, giraffes, bread mold, salmon sperm, and viruses all use the same code. If one removes a nucleotide sequence from a human and places it in the cellular machinery of a frog embryo or a redwood tree, the polypeptide sequence that emerges is the same as the one the human would have produced. It is much like placing a cassette in a tape player; it doesn't matter if the tape is in Russian and you've placed it in a Japanese-made Sony or if it is in English and you place it in a Panasonic made in Taiwan, the message will emerge as intended. The code is universal and the devices for expressing the code all recognize it.

An Interlude: The Role of Conceptualizing in Scientific Thinking

There is a subtleness in the way terms like "code" and "information" slide into the discussion of biological subjects. You may recall some of the 17th century's efforts to comprehend life: ideas borrowed from the combustion processes of candle flames and steam engines. This transfer of a concept from the known to the unknown is not only useful but inescapable. There just isn't any other way to "imagine" an explanation for a totally unknown phenomenon. Our minds work by extension of ideas, and an idea has to have a basis.

"The selection of a conceptual model is critical because it shapes the investigative effort which will follow."

When researchers are faced with an unknown process such as the method by which a gene functions, they must select from among the possible models in their minds. One possibility is that a gene is like a bell and the sound waves emanate from it. Others are that the expression is like that of a stone dropped in a pond with spreading ripples, or a signal fire lit on a mountain top, or the odor of a rose attracting a passing bee. The selection of a conceptual model is critical because it shapes the investigative effort which will follow.

If the genome is perceived to be something like a book then its "information" content is obtained directly by looking at the pages. But if the genome is thought of as a piece of magnetic tape with the genes encoded along its length, then a tape-player must exist which is capable of converting the magnetized locations into sounds which are "meaningful." A person can *look* at magnetic tape forever and obtain absolutely no information from it.

In the case of DNA which is a polymer composed of nucleotides and protein which is a polymer composed of amino acids, the tape analogy is very attractive. Long chains of "information" in the DNA molecule somehow bring about specific sequences of amino acids in the proteins. A linear arrangement of DNA information brings into existence a linear amino acid sequence in the protein. The idea began to form that DNA and protein were **colinear**.

Imagine a string of nucleotide beads. The first of three beads carries the letter A, the second U, and the third G. The first triplet is AUG. The next three beads are UUU, followed by CUA, UUC, and finally UGA. You also have a box filled with amino acid beads. Taking your direction from the nucleotide string, you consult your code book and select the specified amino acid beads and string them in the sequence dictated. Your first amino acid bead is methionine, your second is phenylalanine, your third is leucine, your fourth is also phenylalanine, and your fifth bead is — but wait — is there a fifth bead? No, UGA is a stop codon. Your string is terminated. You tie a knot in your amino acid string and hold it up next to the nucleotide string — AUG,UUU,CUA,UUC on one string, and *colinear* with it is the sequence met, phe, leu, phe and then the knot.

"We've borrowed concepts and used such words as, codes, spaces, and colinearity from the world we know and understand and have extended them into a world we do not know and do not understand but which we hope is comprehensible."

We've borrowed concepts and used such words as codes, spaces, and colinearity from the world we know and understand and have extended them into a world we do not know and do not understand but which we hope is comprehensible. The task for the scientist is to detect, from the kinds of answers Nature gives in response to questions, whether or not, at the molecular level, there are actual structures, processes, and events which correspond to the borrowed concepts.

Concepts and Questions: Some Wrong, Some Right

Some questions asked of Nature yield very informative responses. But asking good questions isn't easy. You cannot ask Nature, "How do you make proteins?" Nature's only responses to scientific questions are "Yes" or "No." The scientist has to be very clever in the construction of the questions.

As soon as Watson and Crick's DNA structure was announced in 1953, the physicist George Gamow suggested that DNA's structure would create, on its outer surface, indentations into which amino acids would fit.

He predicted that there would be differently shaped depressions caused by variations in the nucleotide sequence and that each kind of amino acid would fit into "its" properly shaped slot. We've seen that physicists have been among the most productive contributors to molecular biology. In this case, however, Nature's response was strongly negative. Gamow's suggestion failed on two counts, one chemical and the other biological. The chemistry of DNA's surface and the shape of the 20 kinds of amino acid molecules simply didn't match up. The biological error was of a different sort.

The preponderance of DNA is found in the chromosomes which lie within the nucleus. It seemed clear that the nuclear DNA never left the nucleus. Yet protein formation occurs entirely in the *cytoplasm*. By having the amino acids fitting into slots on the DNA in the nucleus, Gamow's proposal would have had protein formation occurring in the wrong place. But notice how sharply his failed suggestion focuses our attention on the real issue. How *does* the sequence information get from DNA, inside the nucleus, out to the cytoplasm?

Francis Crick had long favored the idea that since DNA remained in the nucleus its information had to be copied in some way permitting the original to remain in the nucleus while the *copy* would be exported to the cytoplasm. He had favored the concept of a **template** as the basis of the copying process. If you have ever seen a paper pattern used to guide the cutting of cloth to be sewn into a garment, you've got the idea of a template.

mRNA

In 1960, Francis Crick, Sydney Brenner, and Jacques Monod met at Cambridge University and were discussing the problem of just how DNA's genetic instructions were expressed. Brenner had been at Oxford at the same time Watson and Crick were at Cambridge. He had a long-standing interest in the problem. It was Brenner who had devised the name "codon." Jacques Monod was at the Pasteur Institute in Paris and together with his long-term collaborator, François Jacob, had performed a series of brilliant experiments which led to their Nobel Prize. Jacob and Monod were convinced that some form of messenger molecule was involved. Reviewing all the experiments that had been performed, these three men decided that the substance which carried the gene's message from the nucleus to the protein synthesizing site in the cytoplasm had to be RNA.

In 1960, Jacob and Brenner spent several months at Cal Tech working with Matthew Meselson. In May of 1961 they published the results of their work in a paper titled, "An Unstable Intermediate Carrying Information From Genes to Ribosomes for Protein Synthesis." Jacob and Monod named the "unstable intermediate" substance **messenger RNA (mRNA)**. We will shortly take up the meaning of the term "ribosome."

How Do You Copy a Gene?

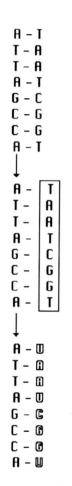

The making of the message starts off exactly like the replication of DNA. An enzyme facilitates the breaking of the hydrogen bonds which connect the two strands of the DNA double helix permitting the strands to unwind and separate. Another enzyme, called **RNA polymerase,** recognizes only one of the two strands. This enzyme moves along the selected strand, and as it encounters each of DNA's **deoxyribonucleotides** it hydrogen-bonds to it a complementary **ribonucleotide.** RNA does not contain the base thymine. In its place it substitutes uracil. So the complementary pairing resulting from the action of RNA polymerase is A-U and G-C.

If the selected DNA strand sequence is A-T-T-A-G-C-C-A, then RNA polymerase would create the complement U-A-A-U-C-G-G-U. The process of using DNA as the template to guide the sequence formation of RNA is called **transcription**. A length of RNA produced this way is called a **primary transcript**. Once the primary transcript is completed, it is released from its DNA template (still another enzyme is involved) and will be readied for transport to the cytoplasm.

Control of Transcription

Following DNA strand separation, the selected strand (unboxed) is used to specify a complementary RNA strand.

Before we send the mRNA molecule on its way, we might as well answer three puzzling questions. How does RNA polymerase make its decisions as to *which* of DNA's two separated strands it will recognize; and exactly *where* along the selected strand will it commence transcription? Also, how is the decision made as to where to terminate transcription?

DNA has regions where its nucleotides form so-called **promoters**. These are locations to which RNA polymerase is strongly attracted. Each gene has its own promotor region adjacent to it. Only one of the two strands can have a promoter sequence at a particular location. (The other strand will have the *complement* of this sequence. A left glove is *not* the same as a right glove; it is its complement.) Along the sequence of nucleotides, there are also regions to which RNA *cannot* bind. These regions function as *terminators* of transcription. In the simplest possible case, the decision as to which genetic region is to be transcribed is made by the **affinity** (chemical binding attraction) of the promotor regions for RNA polymerase.

It is very much like the game you play at a carnival or a State Fair where you roll the ball toward a target which has a bull's-eye arrangement. Getting the ball somewhere inside the outer ring is not much of a challenge; it earns you 5 points and a plastic ring with a spider on it. The inner rings get smaller and smaller until the center ring is barely as big as the ball itself. It is worth 100 points and the teddy bear. There is no law which says you cannot hit the center ring; it is a case of probability. So it is with the control of transcription. The affinity of various promotor regions for RNA polymerase determines the degree to which transcription will take place.

"The affinity of various promotor regions for RNA polymerase determines the degree to which transcription will take place."

There are a variety of protein molecules which can hydrogen-bond to DNA. Some are enzymes like RNA polymerase, and others are **transcription factors** (controlling elements). By binding to various DNA locations, these transcription factors can either decrease or increase the affinity of the promotor regions for RNA polymerase. In this way it is possible to vary the degree of transcription. Decisions as to which genetic regions are turned on and off are the result of the presence of varying kinds and amounts of transcription factors. We will have more to say about this topic shortly.

The Processing of mRNA

All of us have seen copies of the Declaration of Independence, and we take it on faith that the copy we saw was complete and accurate. When a document is extraordinarily important, we assume that great care has been taken in making copies of it. In the early days of mRNA research it was assumed that each precious deoxyribonucleotide in DNA would be meticulously transcribed into its complementary ribonucleotide. And so it seems to be. RNA polymerase is scrupulously precise in making the copies. But after the RNA primary transcript is released from its DNA template, a most disconcerting thing occurs. A series of enzymes cuts the primary transcript, the precious bearer of genetic information, into many smaller pieces. Most of these pieces are then destroyed! Those pieces of the transcript which are spared destruction are then spliced together to make a **processed transcript**.

When RNA-processing was discovered it was absolutely unexpected, and the researchers found it very difficult to believe their results. The destruction of the precious sequence seemed to be an impossibly bizarre event. It was the equivalent of cutting out most of the words in the copies of the Declaration of Independence and then scotch-taping the remaining fragments together. Bizarre or not, that's the way life does it.

The mRNA which emerges from the processing step is **not** colinear with the DNA template. Long stretches of information have been deleted. The genetic message sent down through the generations and preserved in the chromosomal DNA is fragmented, and following the processing what remains is but a small fraction of what had been originally transcribed.

As the knowledge of DNA's contents has increased over the years the depth of this mystery has been revealed. Only a very small portion of an organism's total DNA (in the human about 5%) contains the coded instructions for the production of proteins. The vast majority of DNA is never expressed. Some portion of the remainder plays a role in the control of genetic expression, but there remains an enormous amount of DNA for which we have not been able to find a completely satisfying explanation. At one time some scientists found it entertaining to refer to this unexplained material as "junk" DNA. Fortunately this fad has died out. Not understanding something is one thing; denigrating the unknown is, quite unacceptedly, something else.

"The destruction of the precious sequence seemed to be an impossibly bizarre event. It was the equivalent of cutting out most of the words in the copies of the Declaration of Independence and then Scotch-taping the remaining fragments together."

Coded Messages Need Decoding

If we consider the way in which music is encoded on tape we realize that our ears cannot hear magnetized iron oxide. It is essential to put the tape through a "player," an apparatus which can convert the magnetic code into sound. All coded information must be dealt with this way. When you read a story out loud, you are converting the visual information on the page into sound.

Scientists realized that although they had "broken" the code and learned that UUU meant phenylalanine, they didn't know how the cell actually selected "phe" from among all the other amino acids and how each of the selected amino acids was assembled in the polypeptide chain. This situation is very much like the state of affairs when a stone-age tribe is introduced to a radio. The people accept the fact that a voice is coming out of the box, but there is no comprehension of the mechanism involved. Scientists are particularly unhappy when faced with unexplained realities.

A number of pieces were fitted into place to complete the understanding of the conversion of the mRNA code into a polypeptide chain. The first was the discovery of a molecule which has almost unbelievable qualities.

What Do You Do When the Instructions Aren't There?

"There isn't any-thing more frustrat-ing than learning that the instructions for the assembly of a gift have been thrown out with the Christmas wrap-pings."

Cells contain a variety of different kinds of RNA. Some RNA chains were extremely short, others quite long. Some RNA is so unstable that it breaks down while the researcher is attempting to isolate and study it. Other kinds of RNA are very stable and readily examined. To make things even worse, the various kinds of RNA seemed to make temporary associations with one another. The researchers had a terrible time with RNA.

There isn't anything more frustrating than learning that the instructions for the assembly of a gift have been thrown out with the Christmas wrappings. A jumble of plastic parts, several envelopes full of mysteriously shaped metal fasteners, and the little piece of paper which says, "Inspected by No. 17" are all that remains. There is not the slightest indication of the roles that the various pieces are intended to play. That's what it felt like to the investigators.

If there is no possible way to recover the instructions, solving the problem begins by making guesses as to just what the toy is and how it might work and then identifying, from among all the pieces, likely candidates to fill the various roles. That is essentially how the researchers went about the task of establishing how the various kinds of RNA functioned so as to produce the polypeptide sequences stipulated by the genes.

Biochemists had established that there was absolutely no relationship between the structure of any of the amino acids and their mRNA codons. If one looked at the configuration of the codon UUU there was just no way to make phenylalanine link to it. None of the amino acids related in any chemically sensible way with their codons. We expect relationship in assemblages: this screw is complemented by this nut; this cassette fits into this slot in the tape-player. But we also know how to handle the concept of an *adapter*. When we find that a pair of headphones has a cord and a plug which will not fit into the socket in a transistor, we search for a connecting piece which has the ability to link the otherwise incompatible components. There was a need for some molecule which had the capacity to chemically link mRNA codons with their specified amino acids. From among the many kinds of RNA, one population emerged which played the linking role perfectly.

"There was a need for some molecule which had the capacity to chemically link mRNA codons with their specified amino acids. From among the many kinds of RNA, one population emerged which played the linking role perfectly."

tRNA

Transfer RNA (tRNA) is produced in exactly the same manner as mRNA, by the process of transcription using DNA as its template. There are twenty different DNA template regions so there are twenty different kinds of tRNA which are transcribed. The fact that there are 20 different kinds of amino acids should have rung a bell the minute you read that sentence. We'll return to the number 20 shortly.

If you cut off a piece of Scotch tape and inadvertently allow it to curl back on itself, you've got a mess. You intended a nice straight length of tape but you've ended up with a jumble where the tape has stuck to itself in many places. This is the appearance of tRNA. In a number of locations it has curled back on itself, and A-U and G-C pairings form and hold the originally linear molecule in a tangle. One end of the molecule projects out of the mess, and at another point there is an important loop.

Let's look at the projecting end first. It is a short piece sticking out of the tangle and it is at this end of the tRNA molecule that the amino acid finds a linking location. Each of the 20 different kinds of tRNA has a unique **amino acid attachment site**. For every amino acid there is a tRNA molecule which has an attachment site that will bind that one kind of amino acid and no other. To see exactly how the adapter role of tRNA is played, let's attach phenylalanine to *its* specific tRNA. We now have the adapter and its proper amino acid in position.

Let's be sure we know exactly what we would like the adapter to do. We're trying to get phenylalanine to appear in its proper sequence in a polypeptide chain. The sequence has been stipulated by the string of triplet codons that make up an mRNA strand. The codon for phe is UUU. The task presented to the adapter is to recognize the UUU codon and bring phenylalanine to that location.

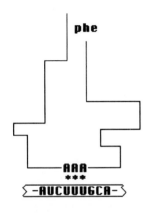

A molecule of tRNA with phenylalanine (phe) at the amino acid attachment site, and the anticodon AAA positioned in the anticodon loop. Below the tRNA is a stretch of mRNA with the codon UUU, which specifies the amino acid phenylananine, paired with the anticodon.

Recall that tRNA has an important and readily identifiable loop. This is called its **anticodon loop**. In the particular tRNA we're examining the ribonucleotides which make up this loop contain the sequence AAA. This triplet of ribonucleotides projects out and is available to form hydrogen bonds. The base-pairing rule now takes over. The mRNA **codon** UUU can form hydrogen bonds only with the tRNA **anticodon** AAA. Carrying phe at its amino acid attachment site, the tRNA's anticodon hydrogen bonds to the mRNA's codon. The adapter has one end carrying the amino acid and its other end has base-paired with the codon which specifies that particular amino acid. The challenge of bringing phe directly to the site where its codon is located has been overcome.

Finally, a Role for the Microsomes

In the early 1960s, when the research reports began detailing the process by which the individual amino acids were brought into their sequence positions, I remember having a set of conflicting emotions. The first was a sense of pride in being a scientist, a member of a profession which was capable of revealing such astonishing phenomena. Hard on the heels of pride, however, came doubt. Could this actually be happening in every cell of every living thing on the planet? Visualizing the mRNA tape with its sequence of codons being recognized by 20 different tRNA adapters each carrying the precise amino acid specified by the codons, I found myself wondering if this molecular mechanism could really exist. As I told my students about this wondrous activity, I looked into their eyes to detect their responses. Did they believe this account? Wasn't this story just too incredible to believe?

"Isolated bits of knowledge that previously had no apparent meaning began forming into a coherent and sensible whole."

There was a third reaction. Finally, after all the years of effort by cytologists, biochemists, and physicists-turned-biologists, all the pieces were falling into place. Isolated bits of knowledge that previously had no apparent meaning began forming into a coherent and sensible whole.

I remembered a course in cytology in which I was urged to enroll by my graduate supervisor. I really didn't want to look at cells under the microscope; my research enthusiasms were more experimental than observational. Mrs. Peterson, the woman who taught me how to prepare cells for observation and how to get the maximum amount of information out of a microscope, had a passion for the tiny dots which were scattered about in the cytoplasm. She called them by the name they had been given ever since their discovery in the 19th century — microsomes, literally "little bodies." If you were a good cytological technician, she said, the microsomes would appear sharply defined, beautifully stained red with the dye acid fuchsin. She could not tell me what the microsomes did. To me, these were little dots without meaning. I am ashamed that I judged my teacher so harshly.

The dots had been renamed, early in the 20th century, when it became clear that they contained RNA and protein. They were called **ribosomes**. There was circumstantial evidence that they were somehow involved in protein synthesis because the ribosomes were always found in those regions where proteins were being formed. In the 1960s they finally came into their own. Mrs. Peterson must have been very proud of them.

Translation

The ribosome may appear tiny under the microscope but it is gigantic in relationship to an amino acid. The molecular weight of a typical amino acid is about 100; the molecular weight of a ribosome is about 4,500,000.

The ribosome is actually a complex assembly of RNA and about 80 proteins. This structure performs the most complex task asked of any cell component. It functions as the tape-player and must properly position the cassette containing the tape; it must move the tape smoothly past the decoding heads; it must transform the coded information into a meaningful signal; and it must eject the tape when the message has been completed. The assembling of amino acids, using the mRNA codon sequence as the template, into a protein is called **translation**.

Ribosomes are standardized. They can recognize and translate any mRNA. A sea urchin ribosome provided with a chimpanzee's mRNA will faithfully produce the chimpanzee polypeptide sequence called for. You may remember that the very first codon is always AUG, which specifies methionine (met). All ribosomes recognize this AUG codon and use it to position the mRNA strand at the **initiation site** on the surface of the ribosome. That signals the "start" of a sequence. Methionine is the first amino acid in what will become the growing polypeptide chain. The ribosome then moves along the mRNA strand one codon at a time, accepting those tRNAs whose anticodons complement the codons it encounters. The amino acids carried by the tRNAs are attached, one by one in sequence, to the growing polypeptide chain. This part of the translation process is called **elongation**.

When any of the three **termination** codons (UAA, UAG, or UGA) is encountered, the ribosome terminates the polypeptide chain by disengaging from the mRNA. The process of translation has produced the polypeptide chain specified by the mRNA's codon sequence, which is quite a job for a tiny cytoplasmic dot.

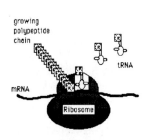

The ribosome (solid), moves along the mRNA molecule. Each tRNA , bearing its amino acid, is recognized by the codon it specifically pairs with. Each amino acid, in turn, is released and is added to the growing polypeptide chain.

The Central Dogma and Beyond

In 1957 Francis Crick presented a succinct message to the scientific world. Life's information flow, from gene to expressed trait, he said, was accomplished by DNA being the template for the formation of RNA, and RNA, in turn, specified the amino acid sequence for the production of protein. This concept, frequently shortened to "DNA makes RNA and RNA makes protein," has been called the central dogma of biology. In many biology textbooks the central dogma is presented as DNA —> RNA —> protein.

This is an interesting use of the term. A dogma is a belief, typically a creed held by a group. Religious convictions are usually what come to mind when the term is used. For scientists to adopt the word in their description of the molecular processes is a little jarring. We do not expect scientists to be "dogmatic." There are some subtleties here which need to be explored.

The arrows are the most interesting part of the central dogma. They state emphatically that DNA is *the* starting point. There is no other way to produce RNA, says the first arrow. The gene is the source of all information. The second arrow, leading to protein, informs us that unless RNA functions as the *intermediary,* the gene's message will not produce protein — so far so good.

Francis Crick had something in addition to this in mind, something which bears directly upon the belief that a blacksmith's son will be born with larger muscles due to his father's occupation. Crick was attempting to put a final nail in the coffin of Lamarck's idea of the inheritance of acquired characteristics.

Notice the *direction* of the arrows. DNA —> RNA —> protein. The blacksmith's muscles or the giraffe's neck are essentially protein. When we lift weights we increase muscle mass; we increase the amount of contractile protein. More significantly, the enzymes which control the metabolic pathways which produce all of the critical substances and activities are protein. So let's show any *acquired* modification in *protein* this way: DNA —> RNA —> **protein**. The boldface type indicates an *acquired* characteristic. But we do not send our muscle protein or our enzymes into the future to our children. We send our genes. In order for the acquired characteristic to become *genetic* there would have to be a corresponding change in the genetic DNA. How would genes be informed of newly acquired larger muscles? There would have to be an informational flow from protein back to RNA and from RNA back to DNA to modify the specific genes responsible for the acquired trait. The diagram would have to be: DNA <—> RNA <—> **protein**.

Now we begin to detect a different emphasis in the central dogma. It is not only a statement that genes send information to RNA and in turn to proteins, but also that there is absolutely no way for protein to send information back so as to modify or change the genes. In other words, any ideas we may have about practicing the piano, or doing good deeds, or learning French so that our children will *inherit* these acquired skills and characteristics come up against the absence of a pathway. Biology just doesn't work that way. *We* can become kinder persons, or skilled linguists, or develop enviable bodies, but our children cannot *inherit* our protein-based acquisitions. Notice the dogmatic sound of that statement.

"It came as a shock to biologists when, in 1970, while investigating a group of viruses, it was discovered that they possessed an enzyme which was capable of forming DNA using RNA as the template; in other words, DNA <— RNA. "

It came as a shock to biologists when, in 1970, while investigating a group of viruses, it was discovered that they possessed an enzyme which was capable of forming DNA using RNA as the template; in other words, DNA <— RNA. This enzyme, called **reverse transcriptase** (it catalyzes **transcription** in reverse), is found in the **retroviruses** whose genes are in the form of RNA. The AIDS virus (HIV) is a retrovirus. We'll consider the implications of this in Chapter Thirteen.

If the arrow between DNA and RNA can be reversed, how about the one between RNA and protein? In spite of decades of intense searching we have never found such a reversal. Protein remains at the end of a one-way information flow. Lamarck cannot find any support from molecular biology.

Then How Do Organisms Change?

Since science concerns itself with providing explanations for the phenomena we observe, it is obligatory that scientists do more than simply state the phenomena in more precise language. When we observe the sudden appearance of a brand new structure or function in an organism (for example, the white-eyed fly in a population of red-eyed *Drosophila*), we call the individual a *mutant,* and we call the process which produced it *muta-*

tion. But that is not an *explanation*. If entirely new structures and capabilities can abruptly appear in organisms, aren't scientists obliged to explain the process?

The nature of mutation was always assumed to be a change in the chemical structure of the gene or genes involved. We've mentioned the use of X-rays to produce mutations, but until the details of the genetic code were worked out it wasn't clear just what the X-rays were doing.

If a **DNA** sequence of deoxyribonucleotides is TAC-AAA-CTT-ACA, then the *complementary* **RNA** transcript of this sequence will be AUG-UUU-GAA-UGU. The transcript's codons stipulate the sequence of the amino acids met-phe-glu-cys. Now suppose an X-ray slams into the DNA and rearranges the third triplet (CTT) so as to form a new, mutated sequence, **CAA**. When this mutant triplet is transcribed it produces the RNA codon **GUU** which specifies the amino acid valine (val). The mutant produces the polypeptide sequence met-phe-**val**-cys. A sudden and new conformation of the polypeptide results. The substitution of **val** for glu may have very significant consequences. I chose this particular mutational substitution because it is the one which causes the disorder **sickle-cell anemia.** The substitution of a single amino acid, valine for glutamic acid, in a polypeptide chain consisting of almost 600 amino acids, changes normal hemoglobin into a form which profoundly modifies both the configuration and functioning of the molecule. The normal hydrogen-bonding pattern is disrupted, and the resulting molecule reacts abnormally leading to severe disability or death.

"Keep in mind that it is the interaction *of a trait with the environment which determines whether that trait will add to or detract from an organism's chances of survival."*

Not all mutational changes are harmful. We must repeatedly emphasize that mutations are *random* changes. It is conceivable that some random changes may produce alternative polypeptide sequences which are advantageous. Keep in mind that it is the *interaction* of a trait with the environment which determines whether that trait will add to or detract from an organism's chances of survival.

The point is, we know the basis of mutation. It is change in the DNA sequence which results in changes in the transcribed RNA and ultimately in the translated polypeptide.

The sources of the energy necessary to effect mutational change are many — various forms of ionizing radiation, certain chemical substances (mutagens), ultraviolet light, among others. As we have emphasized, mutational events occur at random. It is not possible to *direct* a mutation. Even if one knew exactly *where* to aim the energy, it isn't possible to pre-determine the *outcome* of the "hit."

The kind of mutation we've been describing is called a **point** mutation. A very limited change occurs; perhaps one or two nucleotides are involved. There are some mutations which don't affect just one or two nucleotides but may involve losses, additions, or rearrangements of relatively long stretches of deoxyribonucleotides. Obviously this latter kind of change can result in very significant alterations in the traits involved. Once again, however, it must be emphasized that selection forces are the test which all mutations must face.

"So we have come to a point in the explanation of life where we find that at its root all change, all production of variation, is based upon a random modification of the sequence patterns of deoxyribonucleotides."

So we have come to a point in the explanation of life where we find that at its root all change, all production of variation, is based upon a random modification of the sequence patterns of deoxyribonucleotides.

Dr. Weismann's Directions for Making a Baby

There is a debate as to which of life's aspects remains the most elusive in spite of the molecularly based understanding of the "new" biology. Some argue that the brain and its role in creating the mind is easily the least well understood. Others feel that ecological complexity is well beyond our grasp. I champion the view that embryonic development is the ultimate challenge.

For most of the history of biological thought development was dealt with almost entirely in metaphorical terms. Aristotle believed that the process was one in which fluid matter, provided by the female, had form imposed upon it by the male's semen. The clay and potter metaphor was helpful in communicating this **epigenetic** explanation. Epigenesis argues that an external force applies form to the otherwise chaotic matter. In a revolt from the essentially mysterious imposition of form by unknown forces, many scientists of the 16th and 17th century favored the **preformationist** metaphor. They argued that each individual existed as a miniature but essentially complete individual within the egg or, alternatively, the sperm. Development was simply a process of enlargement and the filling in of details. This scheme seems to require that each generation contain within it the preformed miniature of all succeeding generations.

By the close of the 18th century, detailed studies of a wide variety of embryos showed that the preformationist idea was in error. The earliest stages of all embryos clearly did not possess any of the adult structures in miniature. The alternative hypothesis, epigenesis, wasn't particularly satisfying since the supposed formative power remained entirely unknown.

In the closing decades of the 19th century, the study of development began to borrow thoughts from both inheritance and cytology. In 1885, August Weismann, a compelling figure in German biology for nearly 50 years, considered all of the information that had been accumulated in an essay titled, "The Continuity of the Germ-Plasm As the Foundation of a Theory of Heredity." Weismann's suggestion was that the chromosomes in the nucleus contained a substance which consisted of all of the **determinants** (instructions) for the creation of an entire individual. He called the substance the germ plasm. This germ plasm, and its contained determinants, was replicated at mitosis, and so every cell received a copy of the determinant instructions. The most critical cell, of course, was the gamete — the egg or sperm. It clearly had to contain *all* of the determinants, the entire informational content, so as to be able to form the next generation. The problem, for Weismann, was how to explain the formation of all of the various *kinds* of cells. If all the cells had the same determinant content, how did a liver cell become different from a nerve cell or a bone cell?

"If all the cells had the same determinant content, how did a liver cell become different from a nerve cell or a bone cell?"

In a classic example of imaginative intuition, Weismann came up with the solution. The gametes, the germ cells, needed all the information so as to form an entire new individual, but the various kinds of specialized cells required only that *part* of the information which applied to them. In fact, he proposed, that's exactly how a nerve cell *becomes* a nerve cell! It receives only that portion of the instructions which applies to nerve cells! Weismann called his solution the **segregation of nuclear determinants**. Segregation means separation. In Weismann's theory, the nuclear determinants were separated and distributed among the various cell types. The determinants for bone formation went to cells destined to become bone, the determinants for nerve formation to those destined to become nerve. Weismann's theory explained the fundamental difference between **somatic cells** (the specialized cells which compose the bulk of the body such as blood, bone, fat, and nerve), and the **germ cells** (the gametes — eggs and sperm). What could be simpler? The solution was simple, perhaps, but wrong.

All the Cells Have All the Genes

It became clear from extensive studies in the 20th century, that *all* cells, both germ cells and somatic cells, receive the *entire* set of genes. There is no segregation of the determinants of development. Then how in the world can we explain how a nerve cell "knows" which instructions *it* is to use and a blood cell knows *its* portion of the instructions? By the late 1940s developmental biologists were committed to the idea that somehow there had to be a *selective* use of the gene content by somatic cells; the terminology they used was **differential gene expression**. Each cell type expressed its own unique portion of the total genome.

I have frequently stressed the view that well-substantiated information tends to find itself fitting smoothly into patterns of meaning with other pieces of equally valid thought. In fact, the ultimate test of a scientific concept is whether it fits rationally into the already existent structure of scientific understanding.

A short time back we discussed the fact that RNA polymerase, the enzyme responsible for making the RNA copy of a DNA segment, bound to the DNA only at certain **promotor** locations. Only by binding to a promotor region can RNA polymerase transcribe a gene. In addition, we learned that a variety of **transcription factors** function as modifiers of the affinities of the promotor regions for RNA polymerase. If a cell has no transcription factor for a particular promotor, or if the promotor site is blocked, RNA polymerase will not bind to that promotor. The gene associated with that promotor site will not be expressed. It is by controlling RNA polymerase's access to various DNA regions that cells determine which of the genes a particular cell can transcribe. This is the basis of differential gene expression.

Weismann was wrong about giving somatic cells only certain pages of the instruction manual. All cells receive all pages. But in some cells pages 1 to 10 are glued shut. In other cells the book falls open at page 7. Only the gametes receive a copy with all of the pages totally available for expression.

Clearly, the future direction in studying the nature of embryonic development is to understand the sytem which controls the distribution of controllers. We have discovered genes which produce the transcription factors, and we are on the way to understanding why some cells express these genes and others do not. We are once again in the enchanted forest, and once again we are approaching the magic tower. What shape will this one have?

What Is Genetic Engineering?

Engineers are the ultimate problem solvers. They are given an assignment: dam this river; build a bridge that will support a 50-ton load; create a new computer chip which can process data at a particular rate. Genes, we've been told, can be changed by the process of mutation, but mutational changes occur at *random*. How is it possible to engineer a gene if the process of change is not under control?

In the following chapter we will examine those fields of study which revealed an entirely different set of phenomena, collectively termed **recombinant DNA**, which opened the door not only to increased understanding at the molecular level, but to intrusion into that level. Human minds and human intentions have at long last not only seen into the core of the living process but have placed a hand on that core. In much the same way that a human hand first chipped a stone into a knife, that same hand has now gripped the gene and can fashion it into whatever form the human mind can envision.

THE SORCERER'S APPRENTICES

"The basic trouble is that nature is so complex that many quite different theories can go some way to explaining the results."
— Francis Crick

Walt Disney's film *Fantasia* is a remarkable combination of music and visual images. One segment, Paul Dukas's symphonic poem "The Sorcerer's Apprentice," features Mickey Mouse in the role of the young man who serves his master, the sorcerer, by keeping the rather scary-looking laboratory in order. A particularly onerous task is sweeping and mopping the cavernous room in which the sorcerer performs his magic. As the sequence opens, the sorcerer leaves the laboratory and orders Mickey to clean it up. The apprentice decides that the application of a little magic is indicated, and he brings the broom to life and instructs it to sweep the place clean. Things start off well, and a delighted Mickey watches as the broom goes about its work. But the broom becomes too vigorous, filling and spilling buckets of water with such enthusiasm that the laboratory is engulfed. Mickey tries to restore the broom to its ordinary status but does not know the magic words. The broom increases its destructive cleansing, and Mickey, in desperation, seizes an axe and chops the broom into pieces. There is a moment of silence, and then, incredibly, each piece arises and becomes an entire broom. Buckets are multiplied and water cascades like Niagara through the laboratory, sweeping the hapless apprentice down flights of stone steps in a terror-filled flood. The door flies open and the returning sorcerer takes in the scene with outraged eyes. One word, one gesture from the master, and all is restored to order. The genie is back in the bottle.

During the past 50 years the molecular-level explanations of life have led investigators to a view of the living state which no earlier explanatory system of thought could have imagined. We are confronted daily with the actualities and technological potentials of that view. Collectively, we term all of the molecular manipulations, both real and imagined, **genetic engineering.** Mickey, the apprentice, represents uninformed and misguided use of tremendous power. We see ourselves in Mickey, tempted by the possibilities but vulnerable to events the effects of which we may not be able to foresee. We fear the results of activities which we do not fully understand and may not be able to control. The molecular phenomena which made genetic engineering possible were revealed by some of the most remarkable insights human minds have ever experienced.

"We see ourselves in Mickey, tempted by the possibilities but vulnerable to events the effects of which we may not be able to foresee."

Genetic Recombination

One of the convictions that we humans hold most dear is that of personal individuality. Inside our skins we believe ourselves to be islands of uniqueness. Our laws are designed to reinforce that perception: our persons are not to be invaded by others or to be controlled by others. In the era of molecular biology it is particularly discomforting to be told that living things have apparently been exchanging bits and pieces of genetic material since the very beginning of life on earth. If my genome contains fragments of the gene content of other creatures, how much of an individual am I? Do I really know what I mean by *my* identity?

We discovered the process of **genetic recombination** quite early in the study of inheritance. In Chapter Ten the crossing over between chromosomes with the resulting mutual exchange of genetic material was an instance of a general phenomenon whereby genetic material moves from one genetic "container" to another. Researchers learned that the chromosome donated by a male parent exchanges genetic material with its homologous partner donated by the female parent.

We tend to think of our male parent's chromosomes as the essence of *his* genetic individuality just as we view our female parent's chromosomes as the embodiment of *hers*. Finding out that it is a very normal and constant thing for the chromosomes to exchange pieces with one another during the process of forming reproductive cells, to say the least, is unexpected. The integrity which we have assumed for the chromosomes received from each of our parents is an illusion. In the preparation of the reproductive cells (eggs or sperm) which an individual sends into the next generation, the genetic information is shuffled and recombined. As a result, a chromosome in a reproductive cell is a new combination derived from fragments of the original maternal and paternal chromosomes. With each passing generation the genetic content of every pair of homologous chromosomes is reshuffled as maternal and paternal fragments are recombined in ever-increasing variation.

"With each passing generation the genetic content of every pair of homologous chromosomes is reshuffled as maternal and paternal fragments are recombined in ever-increasing variation."

This shuffling process is not only "normal" but is at the core of life's design for survival. Genetic recombination is one of the two fundamental generators of inheritable variation. The other is mutation. Between them these processes assure that life will continually manifest itself in a variety of alternative forms, the only hope it has to escape extinction as it encounters the selective effect of a constantly changing environment.

DNA Can Transfer from One Cell to Another

It was Frederick Griffith (Chapter Eleven) who first demonstrated that a chemical substance which transferred from a dead virulent bacterial cell to a living non-virulent bacterial cell was capable of converting the living cell and its progeny into the virulent form. This was the start of the hunt for the chemical nature of the gene which eventually proved to be DNA.

What Griffith did not know (and neither did anyone else for many years) was that the DNA from one cell could not only be transferred into another but could become **integrated** into the chromosome of its new host. Once installed in its new location, the "foreign" DNA was replicated along with the "host" DNA at each cell division so that the "foreign" genetic traits carried by the DNA fragments were passed down through the generations to all of the offspring. This form of genetic recombination, where fragments of DNA are taken up by a cell, is called **transformation**. The nonvirulent form of the bacteria, for example, was *transformed* into the virulent type.

Microbiologists (those whose specialty is the study of bacteria and viruses) had long known of a peculiar behavior displayed by certain bacteria. Slender filaments (*pili*) would be extended by one bacterium and these threadlike structures would contact and fuse with a neighbor. After a time these filamentous connections would break, and the individuals would go their separate ways. This process was called **conjugation**. It was learned that the ability to extend the pili was a genetic trait; those strains which could do so were designated as F^+, those which could not as F^-. Following conjugation some F^- bacteria were transformed into F^+. From what had been learned from transformation studies it was suspected that genetic material had crossed from one strain to the other through the tubelike filamentous connection. It turned out that this was indeed what was going on.

"Most genetic material is located on chromosomes. But there are very important exceptions, and it was one of these exceptions which was to provide a most important insight which eventually led to the technology of genetic manipulation."

Unlike Griffith's experimental transformation, conjugation and its resulting genetic recombination is typical of the normal processes by which pieces of genetic material are endlessly being transferred from one genetic container to another.

Most genetic material is located on chromosomes. But there are very important exceptions, and it was one of these exceptions which was to provide a most important insight which eventually led to the technology of genetic manipulation.

Bacteria Have a Circular Chromosome and Also Circular Plasmids

Unlike the numerous rodlike chromosomes possessed by most higher organisms, bacteria have a single lengthy circular chromosome. Most, but not all, of their genes are contained in this continuous loop of DNA. In addition there may be smaller circles of DNA called **plasmids**. It is the genes on one of these plasmids (the so-called F plasmid) which determines whether a bacterial cell will be F$^+$ and therefore capable of producing pili. Prior to conjugation, the circular F plasmid may replicate, and one of the copies can break open so as to form a linear strand of DNA. It is this strand which passes through the pili's filamentous connection into its F$^-$ conjugation partner. After successful passage, the strand reforms into a circle and takes up its plasmid existence in its new home. Since the recipient now possesses a copy of the F$^+$ gene, it can form pili and initiate conjugation. Genetic recombination has occurred.

Attention turned to the plasmids for a variety of reasons not least of which was the phenomenon of **antibiotic resistance**. When antibiotics were discovered and the jubilation over their healing capabilities was at its peak, it was learned that there were certain strains of bacteria which were *resistant* to them. Furthermore, the resistant strains seemed to be capable of spreading their resistance to strains which previously had been vulnerable to the antibiotics. During conjugation, antibiotic resistance genes located on certain plasmids (called R, for resistance) could be transferred through the pili to previously non-resistant bacteria. These organisms, by replicating their newly acquired resistance genes along with the rest of their genomes, would thereby pass antibiotic resistance to all of their progeny.

The discoveries involving bacterial conjugation were extremely critical to the future of molecular biology. Bacteria were certainly very handy objects for study. The ease with which they could be raised in their billions in small laboratory culture dishes made them ideal research organisms. They possessed DNA, and they transcribed RNA which eventually was involved in the production of proteins. Their role in disease was clearly important. The concern, however, was that bacterial genes might behave in basically different ways than did the genes in higher organisms. In higher organisms, for example, genes had definitely assigned locations and were lined up on the chromosomes in very specific order. To place full confidence in the genetic information gained from bacteria it would be necessary to establish clearly the way their genetic systems were organized. The plasmids were to provide the critical information.

Do Bacteria Have an Orderly Genome?

There were a number of suggestions as to just how genetic information passed from one bacterium to another during conjugation. An early suggestion was that the donor's entire genome was replicated and the copy was sent through the pili to the recipient organism. The problem with this idea was that the recipient typically didn't display all of the donor's genetic traits. Only some of them appeared to have been transferred. The data showed that the number of genes transferred was variable from one conjugation experiment to another. This was worrisome. Perhaps a bacterium's genes were packaged in some random manner. If so, this was definitely different from the orderly arrangement in higher creatures.

In 1955 François Jacob and Elie Wollman of the Pasteur Institute in Paris wondered if the confusing data concerning the transfer of genetic material might be explained using a very simple assumption. Jacob suggested what he called a "spaghetti" model. Suppose the genes were indeed lined up in precise order and started through the connecting tubule like a strand of spaghetti. Perhaps the entire strand didn't make it all the way through. Perhaps the conjugating bacteria broke the pilus connection before completing the transfer. If so, only the genes at the leading end of the "spaghetti" would be received. Jacob and Wollman put this possibility to the test. They mixed two different strains of bacteria and allowed them to begin conjugation. They then separated samples of the conjugating population at various times. One sample was separated following only five minutes of conjugation, another at ten minutes, and so on. Some samples were allowed to conjugate for an hour. If the genes were arranged in an orderly manner — let's say that gene A was followed by gene B and then gene C, and so on — then the transfer should be directly time-dependent. A

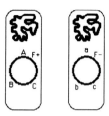

A replicated copy is sent through the pilus.

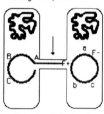

Transfered dominant genes are expressed.

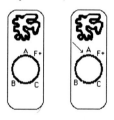

Note that following the transfer, both organisms possess the F$^+$ allele and the dominant A, B, and C alleles.

very brief conjugation might permit only gene A to transfer. A slightly longer one should permit gene A and gene B to transfer. Each increasing time interval should permit the leading end and progressively longer stretches of the following "spaghetti" to make the transfer. The data came out precisely as Jacob and Wollman had predicted. Each circular plasmid had its genes arranged in a precise linear order. The circle always broke at a definite location, and the leading end did indeed start through the pilus with a specific gene at the very front of the line.

You might wonder how the investigators managed to terminate the variously timed conjugations so precisely. After permitting the test population of bacteria to conjugate for the stipulated interval, the organisms were placed in an ordinary kitchen blender and whirled briefly so as to break the pilus connections between them. The blender used was one that Jacob had bought for his wife; being a French cook with some very strong ideas about appropriate kitchen utensils, Mrs. Jacob had refused to use the machine.

The scientific community was assured of the orderly nature of the bacterial genome. Bacteria were entirely valid as genetic model organisms. *Escherichia coli's* genetic system became the most well studied on the planet, and the information gained from this inhabitant of the human intestinal tract was crucial in leading us into what has very accurately been termed the biological revolution.

Bacteria Carry on Chemical Warfare

The viruses which prey upon bacteria, the bacteriophage we discussed in Chapter Eleven, are not unopposed when they enter a bacterial cell. Bacteria produce a variety of enzymes which cleave the invading DNA of the bacteriophage and cut it into a number of small and ineffective pieces. In a similar fashion these same bacterial enzymes can cleave any foreign DNA picked up by the bacteria from their environment. These enzymes are called **restriction** enzymes. It turned out that various species of bacteria produce a wide variety of restriction enzymes. Over 100 different kinds have been isolated and are commercially available.

Just how do these restriction enzymes work? Recall that DNA is typically a very lengthy double-stranded molecule and that its constituent paired nucleotides can occur in any and all conceivable sequences. The following example shows a very short length of an enormously longer DNA molecule. One of the restriction enzymes, *Eco*R I (the first such enzyme derived from the bacterium *Escherichia coli*), recognizes the particular sequence shown and cuts the double-stranded DNA as indicated by the black line.

$$\text{A-T-G}\mid\text{A-A-T-T-C-C-A}$$
$$\text{T-A-C-T-T-A-A}\mid\text{G-G-T}$$

As a result of its action, following the cleavage of DNA by *Eco*R I, the separated ends look like this:

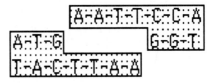

It is apparent that the sequence AATT in the upper strand has been separated from its complementary bases TTAA in the lower strand. Now let's imagine another double strand of DNA from an entirely different organism (shown by the shading) which has also been cut by *Eco*R I:

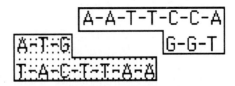

Exactly the same cut has been made by *Eco*R I exposing the same sequence of complementary bases — AATT on one strand and TTAA on the other. Now let's mix the cut pieces from the two different kinds of organisms:

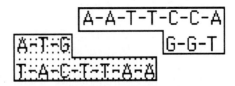

It is clear that the complementary bases AATT and TTAA are in position to bond with one another, and if they do then we will have *joined* the DNA from two different kinds of organisms.

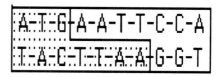

A helpful model for seeing the implications of the functioning of the restriction enzymes is to imagine a set of 100 differently shaped cookie cutters, each one capable of cutting a unique pattern out of a sheet of cookie dough. Imagine using the star-shaped cutter (*Eco*R I) to remove a piece from a sheet of white dough. You have two pieces to keep in mind. The star-shaped piece which will become the cookie as well as the sheet of dough with the star-shaped hole in it. Now imagine a second sheet of cookie dough, this one having been colored green with food dye. Again use the star-shaped cookie cutter to remove a piece of dough. Now you have four pieces to keep in mind. It is possible to put the star-shaped piece from the white dough back into the sheet from which it was cut. It fits perfectly. Press the edges of the cut a bit and the original continuity of the sheet is restored. But it is also possible to put the white star into the green sheet and the green star into the white sheet. After all, you used the same cookie cutter in both cases. Press the edges and you have a green star integrated into the white sheet or a white star integrated into the green sheet. In the case of actual DNA the pieces are joined with one another by another enzyme, DNA ligase.

This is a procedure of enormous power. Quite literally, any and all kinds of DNA can be cut with a particular restriction enzyme. Wherever the enzyme encounters its uniquely specific recognition sequence it will make a cut. Typically such locations are sufficiently numerous that the DNA is cut into many fragments (called **restriction fragments**). Both ends of each restriction fragment are cut at the unique recognition sequence, and therefore *all* DNA cut with a given restriction enzyme yields restriction fragments which can be ligated to *one another*. Pieces of DNA from one species can be inserted into the chromosomes of entirely different species. DNA can be cut into pieces which are reassembled in entirely new patterns. The technological implications are endless. So too are the practical and ethical questions.

From Pure Research to Applied Research to Technology

Bacteria have possessed restriction enzymes for thousands of millions of years. They were part of the protective system which had evolved and which enabled bacteria to destroy foreign nucleic acids which had entered the bacterial cells. Since bacteriophage (viral) DNA had been entering bacteria during this period, it is evident that the restriction enzymes had been producing viral restriction fragments. With processes such as conjugation going on it is inescapable that restriction fragments had been moving back and forth between bacteria for untold periods of time, sometimes being destroyed, sometimes being retained.

During the investigative phase of the work involving plasmids, restriction enzymes, and the various kinds of genetic recombinations, most of us would probably have referred to the research as being "pure." By this we mean that the scientists involved were learning things but they had not established any utilitarian purposes for the knowledge they had gained. For example, it was learned that it was possible to remove plasmids from bacteria for study and then reinsert them into either their original strain or even into totally different bacterial strains. More will be said of this quality shortly.

"A fascinating characteristic of 'pure' scientific investigation is the subtle process by which it is transformed into goal-directed (or 'applied') research. "

A fascinating characteristic of "pure" scientific investigation is the subtle process by which it is transformed into goal-directed (or "applied") research. Imagine the moment when an early human drinking from a quiet pond noticed that the reflecting surface served as a mirror. It doesn't take much imagination for us to envision the next step, when that early human went to the pond specifically to *use* it as a mirror. There is a subtle yet extremely important transition involved. The next step might well have been a decision to put some water in a suitable container, perhaps a depression in a nearby rock, thus saving a trip to the pond. We are justified in using the word "technology" for this example of human control of a natural phenomenon. We can certainly recognize the technological transition that occurred in molecular biology.

Gene Cloning

We have become accustomed to reading about the technological manipulation of genes. When we are told that a gene from a human has been inserted into the genetic apparatus of a mouse we accept the actual transfer as an everyday scientific accomplishment and shift our attention to the societal or ethical implications. The mental image we have of the transfer event itself is of a molecule of DNA from one species having been somehow snipped out of its normal location and inserted elsewhere. This is an oversimplification, and it trivializes some of the most impressive accomplishments of the human intellect.

In order to manipulate genetic material one has to have a great deal of it. One copy of a single gene, literally one segment of a DNA molecule, is simply insufficient for either analysis or manipulation. The molecular biologist needed a way of producing many identical copies of the gene of interest. Restriction enzymes and plasmids provided the means to accomplish this feat of multiplication.

"The plasmids were seen to be ideal devices to serve as genetic replicators. The challenge was to somehow insert the section of DNA containing the desired gene into a plasmid."

Recall that plasmids, like the circular chromosomes of bacteria, are **replicated** when a bacterium reproduces. All of the genes encoded by a plasmid are faithfully reproduced in the copies which pass to the generations of descendants. All of the organisms descended from a single ancestor constitute a **clone**. In theory all of the members of a clone are genetically identical to their ancestor and to one another. We have to qualify this statement because of possible random mutations which may occur. The plasmids were seen to be ideal devices to serve as genetic replicators. The challenge was to somehow insert the section of DNA containing the desired gene into a plasmid. As the plasmid was replicated generation after generation, so too would be the gene. In other words, the gene would be cloned. Huge numbers of identical segments of DNA could be harvested for study and technological use.

The restriction enzymes provided the means for inserting the desired DNA into the plasmid. Recall that it had earlier been learned that plasmids could be removed from bacterial cells and reinserted into bacterial cells. This piece of information would become the basis for a technological technique in which the plasmids would serve as **vectors** (carriers). Plasmids would be removed from their host cells, the genes selected for replication and study would be inserted into these plasmids which would then be replaced into bacterial cells for replication.

Since each restriction enzyme recognizes a specific sequence of nucleotides and makes its cut only at that sequence, the technological problem was to select, from among the many restriction enzymes, one which would recognize a sequence present only once in a plasmid. It turned out that several restriction enzymes fit the job description. Subjecting the plasmids to the cutting action of the selected restriction enzyme resulted in a cut which opened the circular plasmid.

Next came the most ingenious part of the process. The *same* restriction enzyme which was used to open up the plasmid was used to cut the DNA containing the gene of interest. This produced restriction fragments the two ends of which having the same nucleotide sequences at *their* cut ends as are present at the cut ends of the opened plasmid. The DNA fragments were then mixed with the opened plasmids. The complementary ends joined resulting in the insertion of the fragments into the plasmids. DNA ligase was used to seal the pieces together and the desired stretch of DNA became an integrated part of the plasmid. When the bacterial host replicates the plasmid replicated and so did the integrated DNA. This procedure is the basis of **gene cloning**.

More recently a process has been developed in which the selected stretch of DNA can be replicated without involving plasmids or bacterial cells. The **polymerase chain reaction** (PCR) is accomplished using only the gene-containing segment of DNA; a supply of the nucleotides adenine, guanine, cytosine, and thymidine; and the enzymes and chemical reagents necessary for DNA replication. Huge amounts of the desired genetic material can be automatically produced using computer-controlled equipment.

Nucleic Acid Hybridization

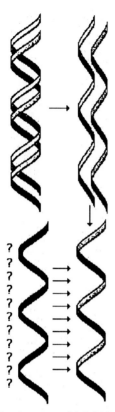

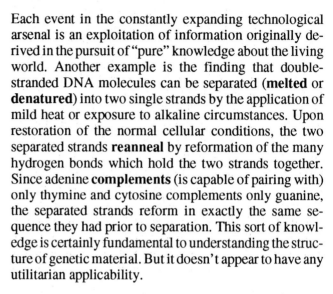

The degree to which DNA from two different organisms can hybridize indicates similarity of sequence. Closely related species have similar DNA sequences.

Each event in the constantly expanding technological arsenal is an exploitation of information originally derived in the pursuit of "pure" knowledge about the living world. Another example is the finding that double-stranded DNA molecules can be separated (**melted** or **denatured**) into two single strands by the application of mild heat or exposure to alkaline circumstances. Upon restoration of the normal cellular conditions, the two separated strands **reanneal** by reformation of the many hydrogen bonds which hold the two strands together. Since adenine **complements** (is capable of pairing with) only thymine and cytosine complements only guanine, the separated strands reform in exactly the same sequence they had prior to separation. This sort of knowledge is certainly fundamental to understanding the structure of genetic material. But it doesn't appear to have any utilitarian applicability.

But it is only a instant of human insight away from an application. What would happen if one mixed separated strands of DNA from two different species of organisms? If the two species had totally different DNA sequences the number of base pairs arranged in complementary sequences by pure chance would be very low, not enough to permit the formation of a **hybrid** mixture of strands. But suppose the two species were quite similar in genetic content and therefore had long stretches of DNA available for recombination **hybridization**? Then, indeed, one would expect to find among the reannealed strands some hybrids representing one strand from each of the two species. A moment's reflection leads to the conclusion that the more similar the two species' DNA sequences are the more hybridized molecules will result. It becomes possible to test for DNA similarity between organisms. In principle, this line of thought underlies much of the testing to establish genetic similarities between DNA from different sources. This technique permits us to ask how genetically similar we humans are to any of the other creatures with whom we share the living state.

We routinely rely upon similarities and differences between samples of DNA in establishing parentage, in probing the question of the genetic differences underlying race, and in identification of criminals from biological fluids or tissue left at the scene of a crime. It took only the blink of an eye to move from what seemed to be purely academic information about molecules to very serious questions concerning the ethical wisdom of probing into the core of our species.

Molecular biology has made possible not only intrusive genetic engineering with its specter of unexpected and unwanted results, but more fundamentally it raises the issue of whether there should be limits to human exploration. We are obviously an incredibly ingenious species. Are there areas of human thought, to say nothing of action, which we should avoid forever? Or, perhaps, until we have evidence of greater wisdom? Or, alternatively, is it a denial of our very essence as thinking creatures to impose limits on thought?

Sequencing the Human Genome

There is probably no more dramatic example of the concern people have over the potential of scientific thought than the **Human Genome Project** (HGP). It is essential to distinguish betweeen **gene mapping** and **DNA sequencing** in order to fully appreciate the potential of this technological Pandora's box.

In Chapter Ten the phenomenon of crossing over was discussed as well as how the use of data derived from this chromosomal behavior was used, by Sturtevant, to "map" the *Drosophila* chromosomes. The use of crossover frequency data has enabled geneticists to map the positions of thousands of genes in all sorts of organisms. We know the positional locations of many of the genes which are clinically significant in humans. Many of the daily media announcements of the "medical breakthrough" variety are reports of the establishment of the locus of a gene whose position was previously uncertain or unknown.

Mapping is dependent upon crossing-over data. In order to detect crossover which occurred in the production of eggs and sperm, it is necessary to observe the effects in the offspring produced by those eggs and sperm. In humans the research is complicated by the fact that humans mate with individuals of their own choosing and

"DNA sequencing is the process of revealing the actual order in which the nucleotides which constitute the DNA molecule occur. There are 3 billion bases in the human genome. In theory, if one can determine the entire sequence all of the coded genetic information is available."

not at the direction of a genetic researcher. The investigator must take the data that the population provides rather than being able to establish a breeding program. Mapping is at best a fragmentary and extremely slow process. Tracing patterns of relationship among family members over the generations is not a project for the impatient. And some genes do not reveal their existence by readily observable expression. One can map only those genes whose existence is known.

DNA sequencing is the process of revealing the actual order in which the nucleotides which constitute the DNA molecule occur. There are 3 billion bases in the human genome. In theory, if one can determine the entire sequence all of the coded genetic information is available. At first glance the process seems laborious but not particularly difficult. Just start at one end of the DNA molecule and work down the nucleotides, one pair at a time, and just keep going. An analogy might be the analysis of the notes making up the songs on a cassette tape. Just start the tape going and keep an account of the sequence of the notes that are played. There is, however, a technological catch. In order to sequence DNA the molecule must first be cut into pieces by restriction enzymes. Our analogy requires that the tape be taken out of the cassette, cut into many, many pieces (which become scrambled in the process), and then each individual piece must be played. This is also not a task for the faint of heart.

In 1989 the scientists who conceived of the project planned that it would be completed in 1995. This sort of massive, long-term effort which involves hundreds of scientists in laboratories all over the world requires a strong and passionate leader. Until he resigned in 1992, James Watson (of Watson and Crick fame) provided that kind of leadership. The budget for this endeavor in 1992 was $164 million. Only a person of enormous self-confidence, international reputation, and intensity of conviction can hope to keep funding at that level flowing from the Congress of the United States. The project continues but not without considerable debate as to its long-term benefits.

What will our tax dollars have bought for us? Theoretically, since all of the genetic information which produces our biological humanity will have been revealed, we will possess the coded statement of what we are and how we are made. We know enough about genomes right now to be a bit cautious about celebrating. The locations in the nucleotide sequence which stipulate specific genes (and ultimately the effects of those genes) are scattered among enormous lengths of nucleotides which do not appear to encode functional genetic information. Most of the DNA doesn't express any genetic message at all. It will take a great deal of time to sift among the sequences, extracting immediately meaningful ones at first, cataloging possible meaning for others, and worrying about the vast majority of the DNA the sequence of which we will know but whose significance we will not.

And along the way there will be fear. This is the ultimate intrusion into the personal space of each of us. The location and content of all of the varieties of humanity's genes will be open for inspection. Each of us can be screened for our secret inner coding — to genetic weaknesses, to predispositions to disease, to malformation, and perhaps to thought patterns and personalities.

Revealed will be not only genes which express physiological products but also those which produce controlling elements. It is these "decision-making" portions of DNA which really lie at the heart of the mechanistic explanation for life.

CHAPTER FOURTEEN

BIOLOGICAL EXPLANATION AT THE MILLENNIUM

"Whatsoever things are true, whatsoever things are honest, whatsoever things are just, whatsoever things are pure, whatsoever things are lovely, whatsoever things are of good report; if there be any virtue, and if there be any praise, think on these things."

— Phillipians 4:8

Have we finally adequately "explained" life? After all the centuries of speculation which invoked vital spirits and humors, inner striving and urges toward perfection, as well as comparisons with flames, steam engines, electrical currents and encoded information, have we, at the close of the 20th century been going in the right direction?

In Chapter One we encountered Jacob's statement:

> For science there are many possible worlds; but the interesting one is the world that exists and has already shown itself to be at work for a long time. Science attempts to confront the possible with the actual. It is the means devised to build a representation of the world that comes ever closer to what we call reality.

It is unworthy of the significance of our question to state, simplistically, that there is always something more to be learned. Of course we will continue to uncover previously unknown pieces of reality. The issue is whether what we may learn in the future will fundamentally alter the explanation. A new species of insect discovered in Brazil has the potential to either fit coherently into our existent explanation of life or to distort it so seriously as to force us to modify our model of reality. Let us permit Jacob to guide us in rendering a judgment about attaining the ultimate biological explanation.

Representations of Reality

Jacob portrays science as building a *representation* of reality. There is a critical distinction between the world "out there" and the representation of it we build within the human intellect. Five thousand years ago the sun appeared and rose in the east and was observed to move across the sky to set in the west. The appearance is exactly the same today. What has changed is our explanation for the appearance. Five thousand years ago the sun was *represented* as moving across the heavens while we on earth stood firmly still. Many years of observation and thought have changed the way we represent the solar system. Now we explain that the earth *revolves* which creates the *appearance* of a sun moving *relative* to us. In creating this representation we use words like revolve, appearance, and relative. These are all human constructs.

A troublesome thought arises when we consider Jacob's contention that the representations of reality that science devises "come ever closer to reality." A skeptic might argue that each successive representation simply differs from its predecessors but isn't really getting closer to reality. I believe that the skeptic is wrong.

Scientific explanations mature in a manner similar to geographical knowledge. When the New World was first encountered by European explorers they had no idea of its dimensions or the structure of its interior. In fact they had no idea if it was the last New World to be encountered or if there were many more as yet unknown lands. But with its discovery one thing was clear; *it* existed. That gave directionality to the following generations of explorers who came directly to it and could extend the knowledge of its coastline and its interior rivers and mountains, and, ultimately, understand its position relative to all the other continents.

The progress of science is similarly both cumulative and directional. We not only add new pieces to the pattern of thought, but, as was pointed out in Chapter One, each new piece must fit coherently. Over time the increasingly convincing pattern points out the direction of future research. For example, the Mendelian proposal that the gene was an actual particle led to the search for its chemical identity. When DNA was identified as the genetic substance we were then challenged to determine its structure and ask if this structure was able to self-replicate and encode information. When these answers came back affirmative we saw that the next logical questions involved the mechanisms by which the encoded messages were expressed. Each step of the way posed a new test for the explanation. On the one hand, if at any point the explanation failed the entire concept was undermined. On the other hand, each newly integrated and supportive finding added conviction to the concept of a genetic particle. We *are* progressively narrowing the gap between the possible and the actual.

"Are there any big surprises waiting, perhaps entire continents of wisdom as yet unseen?"

The Direction of Future Explanation

Is it possible for us, at the close of the second millennium, to project the way biological explanation will progress? Do we have a sufficient understanding at this point in time to make predictions as to what the final map will look like? Are there any big surprises waiting, perhaps entire continents of wisdom as yet unseen?

There are perhaps a half-dozen factors which will affect,

in a major way, the direction and the progress of biological investigation beyond the year 2000. Some of these we have considered as we examined earlier stages in the evolution of biological concepts. They were dominant forces then and will continue to be so in the future. Others have been minor influences in the past, but changing circumstances will make them major forces.

Perspectives Affect the Pathway of Progress

Science is usually portrayed as being impersonal, in other words, independent of the personal attitudes of the scientists. This view of science assumes that since the world "out there" is what it is, all scientists who are properly educated and properly motivated will see it for what it is; reality, after all, is there for all to discover. But science is not impersonal. An example will illustrate my claim.

"This view of science assumes that since the world "out there" is what it is, all scientists who are properly educated and properly motivated will see it for what it is; reality, after all, is there for all to discover. But science is not impersonal."

Animal behavior has been studied, as a scientific discipline, for almost 200 years. During most of that period the scientists involved were almost exclusively male. Behavior was seen through male eyes. Now it might be thought that behavior, like the rest of reality, is what it is, and it would not matter whether behavior is observed by a male as contrasted with a female scientist. True, the behavior is what it is but the observer determines what portions of it to focus upon, which events to record, and how much emphasis to assign various aspects; in other words as we have previously noted, the very first step in observation involves the *selection* of what to observe. When, during the past 20 years or so, women scientists began to be more widely represented in the various areas of behavioral biology, there was a noticeable increase in the number of publications dealing with the relationships between mothers and their offspring and the behaviors of female as contrasted with male members of a population. Further, from the data they were accumulating, women scientists began developing patterns of meaning that their male counterparts had not considered. This points out that there is no such thing as an impersonal progression of scientific thought. The choices as to what to observe and which interpretations to apply are made by the community of scientists reacting to both the evidence itself and the personal inclinations of the individuals involved.

The Darwinian Perception: Gradualism

Another example indicates the way a scientist's personal inclinations can influence the perspective from which that individual examines evidence. Concepts which seemed entirely valid for decades are called into doubt by re-examination of the data guided by insights gained from different perspectives. Darwinian evolution assumed that variations were invariably quite small and that selection between alternatives resulted in very gradual changes in a population over extremely long time periods. The fossil record, in general, does show slight changes occurring over very long time intervals but major changes in the record are often abrupt. Instead of finding a constant gradual transition of one form into another, a distinctly changed type replaces what is assumed to be the ancestral stock. If one has a gradualist conviction, these sudden changes in what should be an imperceptible transition are explained as gaps in the record caused by the failure of the fossilization process to have preserved sufficient numbers of the transition populations. We can understand such gaps. Genealogists who search for their ancestors are frequently frustrated by the absence of a birth record destroyed in a fire or the inability to confirm an immigration date due to a missing page of a ship's passenger list. But there is another possibility which can explain these abrupt changes in the fossil record.

An Alternative View: Punctuated Equilibrium

Stephen J. Gould and Niles Eldredge have proposed that the imperceptibly gradual process favored by Darwin is overshadowed by periodic relatively rapid bursts of change in portions of a population. They believe these "punctuations," as they named them, are the major influence in evolutionary change. While not denying that some modifications are gradual, they argue that populations remain relatively unchanged for most of their existence (periods of equilibrium) punctuated by much more rapid, relatively brief bursts of major modifications. If you think about the two alternative explanations you realize that the fossil record is, indeed, what it is, and it is the task of the scientist to *interpret* that record. It has been suggested by those who disagree with Gould and

Eldredge that their views are flavored by their personal preferences, that their ideological position draws them to see sudden and abrupt changes (revolutions) as the essential driving forces both in human history and in natural phenomena. Of course the argument can be reversed. Gradualism as an explanation may also have been motivated by an ideological preference.

"If we are to make a rational prediction about the direction of future biological explanations we should keep in mind that data must be interpreted by people."

Intellectual preferences or predispositions are seen as natural and even essential in politics, religion, and the arts. We feel quite uncomfortable, however, when evidence of personality intrudes into scientific conversations. To a large degree the responsibility for the general belief that science is impersonal must be assumed by scientists themselves. They have fostered the image of science as a process which has built-in safeguards against opinions and preferences. If we are to make a rational prediction about the direction of future biological explanations we should keep in mind that data must be interpreted by people.

Science as Servant

Shortly after World War II, the National Science Foundation (NSF) was created as this nation's basic science forum and source of support. One of the founding principles was that the support and dissemination of science in its purest form, the search for meaning and pattern in the universe, was the primary (and some felt, the only) role of the NSF. For most of the ensuing half-century there has never been much dispute about the proper mission of this governmental agency, but recently there has been a demand from a number of quarters that the mission be modified in the light of restricted resources and a changing sense of the proper role of science.

Proponents of a change suggest that science should serve a much more practical role in the service of society. This point of view maintains that pure (unapplied) investigations which seek to explain the nature of reality are much too expensive in terms of talent and funding and do not contribute sufficiently to the technological needs of the nation. This view perceives the proper role of government-supported science as the provision of the knowledge which assures material progress. The assumption behind this view is that science can accomplish just about anything demanded of it. This is a reasonable conclusion since our century has seen the proliferation of more technological advance, as the result of new scientific understanding of Nature, than has occurred in all the rest of human history.

At first glance this seems fairly reasonable; shouldn't practical benefits in health, agriculture, and ecology be expected from the research efforts of biologists? But think back a bit and ask yourself how much financial support would have been provided to Charles Darwin to spend his life pondering how the living world came to be the way it is. Fortunately for biology, Darwin was an independently wealthy man. Perhaps you may believe that we could have done just fine without an evolutionary perception; so let's turn to Gregor Mendel. Would you have approved a governmental grant to support his work on garden peas? As an agricultural official, would you have seen the technological possibilities of his work? How about Miescher's investigations on the chemical content of salmon sperm? At the time he was pursuing his studies there was no possible way he or anyone else could have predicted that DNA would become the most studied molecule in human history. It is easy in hindsight to see the value of the virus which preys upon *E. coli* but the phage geneticists would have had a very difficult time proving to NSF that their findings would lead to bioengineering in agriculture and medicine. The practical value of science is very easy to see in hindsight but time doesn't run backward.

Let's be practical, however. Are all scientific investigations equally worthy of support? Isn't society obligated to be prudent in its use of funding? How might we arrive at a reasonable distribution of goal-directed as contrasted with pure research? Who should make the decisions? We've encountered another complicating factor in our attempt to envision the future of biological thought.

A Loss of Innocence

For most of its history biologists themselves determined the proper subject matter and the conduct of their investigations. The passion to know, to explore, to explain drove the investigators along pathways that they felt would lead to increasingly accurate representations of reality. The impulse to understand reality was overwhelming and tended to hold in check the human impulse for personal glory. There were a few incidents of serious scientific misbehavior, but these were sufficiently rare as to be overlooked as the result of a disturbed personality. But in the past decade the incidence of intentional deception in the publication of scientific findings has increased alarmingly and has been called to public attention by the media. In our effort to be predictive about future explanations we must factor in this disturbing reality.

"These are not oversights or errors of judgment. They are certainly not scientifically valid disputes as to the best interpretation of data. These are outright acts of fraud."

The departures from traditional behavior have involved absolute fabrication of data, the withholding of results detrimental to a preferred conclusion, and the appropriation of another's results so as to claim precedence of discovery. These are not oversights or errors of judgment. They are certainly not scientifically valid disputes as to the best interpretation of data. These are outright acts of fraud.

Science has become sufficiently enmeshed in the affairs of the world that the world's rewards, be they fame or fortune, are too strong for some scientists to resist. In the face of temptation they steal. Theft from the process of building an ever-progressing representation of reality may not seem as serious a crime as stealing from a bank, but a moment's reflection will reveal the potential for significant hurt to society. A fabricated finding leads other scientists astray, causing them to follow a false trail. Time, effort, and funding are all wasted. But the greatest loss is in confidence in the process. People execute scientific research. The process is totally dependent upon the integrity of its practitioners. Scientists become progressively less secure in their confidence in one another with each revelation of fraudulent behavior.

The various scientific agencies which provide research funding have established investigative processes which examine charges of scientific misconduct and educational, and research institutions have put into place committees of scientists to review the practices of their peers. It is quite evident that science would prefer to supervise itself. But there have been increasingly insistent calls for governmental supervision. The public is concerned that science may not be able to discipline itself. It is unclear what this situation implies for the future conduct of research. It makes predicting much less certain.

Are All Biological Questions Appropriate?

Human beings have always wondered about differences. Why is one racehorse faster than others? How come the offspring of such a horse are frequently faster than average? How much of a role is played by environment as contrasted with inheritance? In hunting dogs and human musicianship we ponder the relative contributions of genes and training in the production of excellence. Are fundamental patterns of growth, function, and behavior basically innate, or are they environmentally modifiable? Some behaviors are very difficult to modify regardless of the environmental circumstances. In others the environment seems to be capable of extensive modification. In 1975, E. O. Wilson published a book, *Sociobiology: The New Synthesis*, which attempted to place the question of genetic and environmental influences in perspective by examining an enormous range of behaviors in a variety of organisms. The title is provocative since it unifies social and biological aspects of the living state. The content is even more provocative since it examines human behavior with the same set of assumptions used for the other organisms. Sociobiology hit the covers of news magazines and was the topic of choice for campus debates. Not since Darwin had there been such an outcry. Wilson's critics were outraged that he should extrapolate from the innate behaviors of lower creatures to the social interactions of his fellow human beings. Biology, it was claimed, had intruded where its findings and research processes were totally incapable of dealing with the complexities. The critics proclaimed that culture is uniquely human and no insights gained from other animals or plants are appropriately transferable. Intelligence, they argued, is not to be explained like eye color. They demanded that the ghost of racial differences in talent or potential not be given credibility as a subject for scientific discussion.

"The critics proclaimed that culture is uniquely human and no insights gained from other animals or plants are appropriately transferable. Intelligence, they argued, is not to be explained like eye color. They demanded that the ghost of racial differences in talent or potential must not be given credibility as a subject for scientific discussion."

Wilson and those biologists who supported him countered that for human beings to willfully suppress understanding of any kind is to deny themselves as *Homo sapiens* (sapient = wise). Biology forced a confrontation between the capacity to question and the prudence of doing so. Are there areas of biological explanation which are seen as socially disruptive to the point where they will not be supported or perhaps declared illegal? If thinking can be seen as potentially destructive, how much more dangerous is doing?

To Engineer or Not to Engineer, That Is the Question

Once an explanation has come so close to reality that its use permits technological advance, we tend to lose sight of the thought process and focus attention on the products. When theories concerning electromagnetic radiation led to the development of radio and later television, we found ourselves glued to the product. Nobody debates the accuracy of the theory now that color screens are wall-sized. The debate is whether or not there is too much sex and violence portrayed during prime time.

The same situation developed with the series of discoveries in nucleic acid structure and the development of the concepts of transcriptional and translational controls. We had approached reality so closely in these areas that we were in the position to develop technological products — designer genes.

This intrusion into what had previously been one of Nature's most secret places has led to very predictable reactions. Those with an enthusiasm for technology (either as suppliers or consumers) see bioengineering as essentially good. It holds out the promise for agricultural improvements, advances in health, industrial applications, and, perhaps, a final understanding of the processes which underlie human thought itself.

In contrast, those who distrust technology express concern about the unknown consequences of releasing genetically modified organisms into the environment and point out that every technological advance in human history has generated unforseen problems ranging from disturbance of ecological systems to massive economic disruption. The term "playing God" is used, by these opponents, to describe what bioengineering intends when it describes the production of organisms modified by genetic manipulation. The researchers and commercial organizations involved in genetic recombination point out that Nature, through the processes described in Chapter Thirteen, has been shuffling bits and pieces of DNA between organisms since life came into existence and that natural genetic recombination is an integral part of the living state. Opponents respond that recombination which occurs naturally is one thing and is not to be confused with intentional manipulations. These, it is stressed, are the result of human intentions. The opponents of biotechnology say that they doubt the judgment, prudence, and responsibility of those who will design and control the recombination processes. They are not convinced that anyone is able to assume such awesome responsibilities.

The Possible and the Probable

Is recombinant DNA technology as potentially powerful as it has been painted? Can we manipulate genes so as to inadvertently produce the radish that ate Philadelphia? A thoughtful response is that given enough time and effort, even with our present state of understanding, it is certainly possible to transfer genes between species, to control the expression of genes, to add and subtract genes, and to combine them in new patterns. But it is important to recall what it is that genes actually do: by their transcription and translation they yield either RNA or protein as their final product. These products must then function within the complex of other chemical products that make up a functional cell and that cell, in association with billions of others, must form an integrated organism. Imagine a human embryo whose genetic constitution is deficient for the production of a critical product, a hormone let's say. If the rest of the organism is normal the insertion of a gene for the production of the missing hormone would permit the entire system to function. It is something like replacing a defective spark plug in an otherwise functional car. Gene therapy can certainly be effective where the problem is a missing substance. The prospects for such restorative therapy are excellent.

Certain birth defects are the result of failures in the *coordination* of gene activities. These processes are much more complicated not only because they involve a number of gene products but, additionally, because the gene products control the production of one another. Simply adding a missing component will not correct things, and even adding all of the components will not restore normality. In these cases we will have to learn the precise timing and placement of the components. Embryonic development remains one of the most poorly explained aspects of life.

A 21st-Century Bestiary?

We will undoubtedly be able to modify certain aspects of existent species. Slowing down the ripening of bananas so they won't turn brown the minute you bring them home from the store is feasible, but having banana trees grow and survive in Minnesota is probably not. Nor is the production of entirely new forms of life feasible.

"In our day this instinct to make ourselves uneasy about the hidden forces in life is manifesting itself again as we imagine vegetables getting out of control and turning on their cultivators or long-extinct dinosaurs being reconstituted from their fossilized genes."

We, as a race, seem to be fascinated by combinatorial monsters — horses with wings, women with the tail of a fish, and men with the bodies of horses. Such creatures fill our myths. Various cultures accepted as absolutely reasonable tales of men with the heads of dogs and trees which gave fruit in the form of children. Medieval bestiaries were collections of descriptions and illustrations of remarkable combinations of various portions of animals and plants. The idea that it is reasonable to combine various manifestations of life into a single organism so as to produce either a horror or a divinity appeals to our sense of the possible. In our day this instinct to make ourselves uneasy about the hidden forces in life is manifesting itself again as we imagine vegetables getting out of control and turning on their cultivators or long-extinct dinosaurs being reconstituted from their fossilized genes.

Life Has a History Which Limits Its Future Possibilities

It is always a mistake to deny the possibility of an event, but life has certain qualities which make some things much less likely to occur than others. Let's consider forming a third eye in the back of the head. It would certainly be useful for marathon runners who would no longer have to turn around to see whether they are being overtaken. We have to understand that we do not have a gene for eye formation, or to put it more emphatically, we do not have two genes, one for left eye and another for right eye. An entire constellation of genetic events must be coordinated to produce an eye. So this is not a case of simply inserting a third gene. In an embryo eyes begin by the formation of two outgrowths from the primitive brain, one on the right and one on the left. The rest of the brain has other assignments, and if we were to provoke some region into producing a third outgrowth we'd end up missing that region's normal function. But that's only part of the problem. Each eye has specific portions of the brain with which it must make neural connections. We really see with our brains. Over the millions of years of evolution the various brain regions have established their various connections with ears, noses, taste buds, and the like, to say nothing of all the special activities associated with integrating various stimuli, reacting to environmental circumstances, remembering, and learning.

"Life is not like a loose-leaf notebook whose pages can be shuffled into any arbitrary sequence. It has reached its state of incredibly balanced forces so that any attempted change faces the overwhelming challenge of simultaneously making a modification without collapsing the entire construction."

Life has a history. It has gone through the process of evolving its reality over the billions of years of its existence on this planet. There are events which occur in an embryo today which have survived and been added to and modified through time beyond imagining. The creation of the genetic recipe for a living thing is rooted in uncountable failures and extinctions which revealed the successful path. Life is not like a loose-leaf notebook whose pages can be shuffled into any arbitrary sequence. It has reached its state of incredibly balanced forces so that any attempted change faces the overwhelming challenge of simultaneously making a modification without collapsing the entire construction. If you want an eye in the back of the head you'll have to trace your way back through the eons past the dinosaurs and the armored fish to the primitive creature which first attempted eye formation and then thread your way back through a totally revised brain operating a totally revised body.

For life, past is indeed prologue. It shouldn't sadden us to realize that not every imagining is within life's design. As we look back at what really has lived and is now living there are marvels and wonders sufficient to give evidence of the possibilities as yet unexplored. The task of explaining these will require imaginings sufficient to challenge the scientists of the future and give them the same joy of accomplishment that has rewarded the efforts of their predecessors.

Index

A

Activation energy, 151
  barrier, 152, 153
Adaptations, 92
Adenine, 187
Agriculture, 164
AIDS
  reverse transcription in, 234
Air
  in Greek thought, 22
Allele, 166-168
  dominant and recessive, 168
  segregation of, 167
Alloway, James, 193
Altmann, Richard, 187
Amino acid sequences
  specificity of, 190
Amino acids, 137
  use in genetic code work, 220
Analogies, 44
Anatomy, 21
Anaximander, 36, 37
Animal behavior, 260
Antibiotic resistance, 245
  role of plasmids in, 245
Aristotle, 17, 31, 35
Artificial selection, 77, 96
Atomic weight, 122
Atomism, 121
Atomists, 32
Atoms, 106
  Greek hypotheses, 121
Avery, Oswald T., 193

B

Babylonians, 11, 12
Bacon, Francis, 139
Bacteria, 193
Bacteriophage, 197
Bateson, William, 159, 174
Beadle, George, 215
"Beads-on-a-string" model, 179
Beagle, The, 79
Behavior, 260
Behavioral biology
  male and female perceptions, 260
Bernard, Claude, 103, 113, 185
Bestiaries, 268
Biology, 105
Biosphere, 146

Birth defects
  genetic engineering and, 267
Blood, 22
Blueprint, 40
Boivin, André, 198
Bonnet, Charles, 117
Bragg, Lawrence, 202
Brenner, Sydney, 224
  mRNA work, 224
Bronowski, J., 1
Buffon, G. L., 39, 56, 60, 162
Butschli, Otto, 172

C

Calorimeter, 5
Cambridge University
  Cavendish laboratory, 202
Capillaries, 28
Carbohydrates, 106, 186
Castle, William E., 176
Catalysis, 153
Catalysts, 153
Catastrophism, 56, 66
Cavendish Laboratory, 202
Cell membranes, 135
Cell theory, 110, 111
Central dogma, 232
Chance, 72, 73
  and necessity, 101
Chargaff, Erwin, 199
  base-pairing rules, 199
Chase, Martha, 198
Chemical bond, 123
Chiasma, 179
Christian doctrine
  Creation and, 32
Chromosomal basis of inheritance, 174
Chromosomal chiasmata, 179
Chromosomes, 172, 173, 177, 178, 180
  bacterial, 245
  mapping of, 254
Classification, 17, 42
Clone, 251
Code, genetic, 219
Coded messages, decoding required, 228
Codon
  defined, 221
  specification of amino acids, 221
  "stop," 221
Cogito, ergo sum, 37
Cold Spring Harbor, 203

Colinearity
   of DNA and protein, 223
   modification of, 227
Columbia University, 176
Communal activity, 8
Comparative approach, 65
Comparison,
   as essence of experimentation, 6
Competition, 89, 91
Complementarity,
   of enzyme-substrate, 154
Concept formation, 6
Conjugation, 246
   bacterial, 244
   in genetic recombination, 244
Consistency, internal, 9
Constancy of organization
   life's maintenance of, 115
Copernicus, Nicolaus, 21
Correns, C., 166
Covalent bonds, 125
Creativity, 136
Crick, Francis, 200, 224, 241
Cross, genetic, 168
Crossing over, 179
Crossover frequency, 180
Culture, effect on science, 5
Cuvier, Georges, 64, 66
Cytoblastema, 112
Cytologists, 172
Cytoplasm, 111
   as site of protein synthesis, 224
Cytosine, 187

D

Dalton, John, 122
Darwin, Charles, 31, 73, 74, 75, 87
Darwin-Wallace explanation, 95
Darwin's finches, 73
Dawkins, Richard, 69
De Humani Corporis Fabrica, 21, 23
De Motu Cordis, 25, 27
De Rerum Natura, 32
Decoding, of coded messages, 228
Deduction, 75
Delbrück, Max, 197-198
Democritus, 28, 32, 39, 72, 73, 119
Deoxyribonucleotides, in RNA produc-
   tion, 225
Deoxyribose nucleic acid, 187
Descartes, René, 37

Development, 236
DeVries, H., 166
Differentiation, Weismann's explanation
   for, 237
Diploid, 198
Disorder and entropy, 147
Dissection, 21
Diversity, 17
DNA, 187, 188, 209
   application of Chargaff's rule, 208
   chemical composition, 187
   complementary strands, 209
   helical configuration of, 205
   hybridization in criminal cases, 253
   model building, 207
   number of bases in, 255
   purine-pyrimidine pairing, 208
   recombinant, 239
   replication, 217
   role in mutation, 234
   role of hydrogen bonds, 209
   self-replication of, 201
   tetranucleotide structure for, 188
   use in establishing relationships, 253
DNA ligase, use in technology, 249
DNA/protein mixture, separation by enzyme
   treatment, 195
DNase, 196
Dominance, 166, 168
Double Helix, The, 207
Drosophila melanogaster, 175
   as genetic model, 182

E

E. coli, 197
   as genetic model, 221
Eddington, Arthur, 147
Eggs, 198
Eighth Day of Creation, The, 204
Einstein, Albert, 6, 110
   concepts precede perception, 6
Eiseley, Loren, 183
Eldredge, Niles, 261
Electron, 123, 124
Elements, 106
Embryology, 175
Embryonic development, evolutionary
   limitations imposed, 269
Emergent properties, 132
Empedocles, 29
Energy, levels of, 124, 140

Entropy, 144, 147
Enzymes, 154
Epigenesis, 236
Equilibrium, 143
Erosion, 56
Error, 10
Escherichia coli, 197
    as genetic model, 221
    in recombination studies, 247-249, 263
Evolution, 36, 99
Exergonic and endergonic reactions, 148, 150
Experiment, 7
Explanation, 1

F

F_1 generation, 167
F_2 generation, 168
Fabricius, Heironymus, 25
Fertilization, 198
Filials, 167
First Law of Thermodynamics, 142
Fitness, 89, 92
  definition of, 91
Flemming, Walther, 173
"Fly room," 176
Fol, Hermann, 172
Forms, Platonic, 31
Fossil, 47, 64
Fossil record, 61
    gradualist interpretation of, 261
    punctuated equilibrium interpretation
      of, 261
Fox, Sidney, 137
Franklin, Rosalind, 204, 211
Fruit fly, 175

G

Galapagos Islands, 73
Galen, 21, 22
Galileo, 33
Gamete, 172, 173
Gamow, George, 223
Garden pea, 165
Gene, 166, 181
    and production of traits, 181
    cloning, use of plasmids in, 251
    differential expression of, 238
Generation, 160
Genesis, 32, 37
Genetic code, 40

degeneracy of, 219
    universality of, 221
Genetic engineering, 239, 242
    politics of, 266
    recombinant DNA in, 239
Genetic notation, 166
Genetic particles, 162
Genetic recombination, 242
    as natural phenomenon, 242
    fear of, 266
Genetics, 161
Genome, 215
    metaphors and models of, 222
Genotype, 168
Germ cells, 198
    continuity of, 238
Germ plasm theory, 237
God, proof of existence, 71
Gravity, 41
Great Chain of Being, 35, 45, 46
Great Flood, 47
Griffith, Frederick, 190, 244
Guanine, 187

H

Haploid, 198
Harvard University, 176
Harvey, William, 25
Heart, 22
Heisenberg, Werner, 2, 75, 206
Hershey, Alfred, 198
Hertwig, Oskar, 173
Heterozygous condition, 170
Hippocrates, 30
Hitler, Adolph, 197
HIV, retroviral nature of, 234
Homologous structures, 45
Homologous chromosomes, 243
    genetic recombination in, 243
Homology, 44–46
Homozygous condition, 170
Hooke, Robert, 110
Human Genome Project, 254
Human genome sequencing, 254
Humors, 29
Hutton, James, 58, 60, 67
Hybrid, 163
Hybridization, 163
    of nucleic acids, 252
Hydrogen, 122
Hydrogen bonds, 129, 205

Hydrophilic polar molecules, 133
Hydrophobic non-polar molecules, 133
Hypothesis, 8, 9
Hypothetico-deductive method, 114
Hypothetico-deductive reasoning, 75-78

I

Ideals, 31
Ideological influences, 262
"If, then" reasoning, 76
Incomplete dominance, 170
Independent assortment, 171, 174, 176
Induction, 74
Industrial Revolution, 5
Information
    DNA as carrier of, 222
Informational molecules, 117
Inheritance, 160
  chromosomal basis of, 174
  of acquired characteristics, 53, 80
Interior mold, 40, 116
Internal consistency, 9
Internal environment, 115
"Invisible hand," 81
Ionic bond, 125, 129
Ionizing radiation, 176
Ions, 129

J

Jacob, François, 4, 7, 51, 224, 246, 258
    mRNA research, 224
    "spaghetti" model, 246
Janssens, F. A., 179
Johannsen, Wilhelm, 167
Judson, Horace Freeland, 204

K

Kierkegaard, Soren, 15
King's College, 204

L

Lamarck, Jean Baptiste, 53
Lamarckian inheritance, 160
Lamarckism, 53
Lavoisier, Antoine, 5, 77
Law of supply and demand, 83
Laws of Thermodynamics, 141
Layered strata, 57
Levels of organization, 108

Levene, P. A., 188, 194
Life force, Lamarckian, 54
Linear arrangement of genes, 179
Linkage, 175–176
Linked genes, 177
Linnaeus, Carolus, 42
Linne, Carl (*see* Linnaeus), 42
Lipids, 106, 133, 186
Liver, 22
Lucretius, 32, 39
Luria, Salvador, 197

M

MacLeod, Colin M., 194
Malthus, Thomas R., 83
Mammals, 48
Mapping, chromosome, 254
Matthaei, Heinrich, 219
Maturity in science, 185
Maupertuis, Louis Pierre, 117
McCarty, Maclyn, 194
Medewar, Peter B., 213
Medicine, 20
Medieval, 33
Medieval scholasticism, 30
Mendel, Gregor, 97
Mendeleev, Dmitri, 122
Mendelian factors, 166
Mendel's First Law, 167
Mendel's Second Law, 171
Meselson, Matthew, 217
    mRNA research, 225
Messenger RNA, 225
Metabolic pathways, 156
    mutant modifications of, 216
Metabolism, 156
Metaphor, 6, 36, 46
Method of questioning, 75
Mice, in pneumococcal research, 191
Mickey Mouse, as sorcerer's apprentice, 241
Microscope, 110
Microsomes, name change, 230
Middle Ages, 31
Miescher, Johann F., 186
Milieu intérieur, 113
Miller, Stanley, 137
Minimal medium, used to detect mutants, 216
Mirsky, Alfred E., 195
Models, 36
    in creating explanations, 40
Molecular genetics, 165

Molecular weight, 122
Monod, Jacques, 224
  mRNA research, 224
Monomers, 187
Montaigne, Michel, 119
Moravia, 163
Morgan, Thomas Hunt, 175-180
Mountains, 57
mRNA, 224, 225
  processing of, 226, 227
Muller, Herman J., 175, 203
Museums, 18
Mutagens, 236
Mutant gene, 176
Mutation, 175
  DNA sequence changes in, 235
  X-rays and, 216

N

National Science Foundation, 262
Natural history, 16, 42
Natural philosophy, 28
Natural selection, 90, 93, 95, 97
Nazi persecutions, 197
Neurospora crassa, 215
Newton, Isaac, 33, 60
Newtonian mechanism, 38
Nirenberg, Marshall, 219
Nitrogenous bases, 187
Nobel Prize for DNA structure, 211
Nomenclature, 17
Nonpolar covalent bonds, 125
NSF, 262
Nuclear determinants, 237
Nucleic acid, 106, 186
  denaturation of, 253
  hybridization of, 252
Nuclein, 186
Nucleotides, 187
  in genetic code, 219
Nucleus, 111, 188

O

Objectivity, 3
Observations, scientific, 2
Occam's Razor, 123
On the Origin of Species, 74
One gene, one enzyme hypothesis,
  214
Oparin, A. I., 136

Orbital region, 123
Order and harmony, 29
Organic chemistry, 106
Organization, 107
  layers of, 108
  pattern of, 45
Origin of life, 136
Oxygen, 122

P

Paleontology, 65
Paley, William, 70, 79
Pangenesis, 162
Parental generation, 167
Particulate inheritance, 164
Patterns of meaning, 29
Pauling, Linus, 204, 207
Pauling, Peter, 207
PCR, 252
Pea plants, 165
Periodic table, 122
Perutz, Max, 202
Phage group, 198
Phenotype, 169
Phenotypic expression, 169
Phillipians 4:8, 257
Philosophie zoologique, 53
Phospholipid bilayer, 134
Phospholipid monolayer, 134
Phospholipids, 134
Pili, in conjugation, 244
Pisum sativum, 165
Plasmids, 245
  as vectors, 252
  in technology, 250
Plato, 31
Playfair, John, 59, 60
Pneuma, 22
Pneumococcus, 191
Pneumonia, 190
Point mutation, 236
Polar covalent bonds, 125
Polar molecules,
  hydrophilic, 133
  hydrophobic, 133
  importance of, 127
poly U, 220
Polymer, 187
Polymerase chain reaction, 252
Polypeptides, 220

Population, 88
  dynamics, 83
  potential, 83
Potential energy, 143
Predictability, 165
Prediction, 11
Preformation, 236
Primary transcript, in RNA production, 225
Principles of Geology, 79
Probability in genetics, 164
Promotors, in differential gene expres-
  sion, 238
Proteins, 106, 186
  as informational molecules, 189
  specificity of, 190
Punctuated equilibrium, 261
Punnett, R. C., 174
Punnett Square, 168
Purines, 200
Pyrimidines, 200

R

Racism, fear of, 265
Randomness
  in variations, 98
  restrictions on, 131
Ratios, genetic, 169
Ray, John, 35
Reality, representing, 4, 258
Recessive alleles, 168
Recombinant DNA, 239
  technology possibilities for, 267
Recombinant genetics, 165
Recombination, genetic, 180
Renaissance, 30
Research, pure and applied compared, 250
Restriction enzymes, 247
  as defense in bacteria, 247
  description of action, 249
  recognition sites for, 252
  use in technology, 249
Restriction fragment, 249
Retroviruses, 234
Reverse transcriptase, 234
Reverse transcription, 234
Ribonucleic acid, 187
Ribose, 187
Ribosomes, 231
RNA, 187
  anticodon loop of, 229
  cytoplasmic location of, 214

difficulty in assigning function, 228
  implicated in protein formation, 214
RNA polymerase, 225
  in differential gene expression, 238
RNase, 196
Rockefeller University (Institute), 188
Rocks, 57

S

Salmon sperm, 186
Schleiden, Matthias, 110
Schneider, A., 172
Schrödinger, Erwin, 202
Schwann, Theodor, 110
Science
  evolution of, 10
  experimentation in, 6
  fraud in, 264
  government support of, 262
  journals of, 8
  observations on, 2
  limitations of, 13, 268
  public expectations of, 263
  questions in, 3
  universality of, 9
Scientists, women, 260
Second Law of Thermodynamics, 143
Sediments, 63
Segregation, 173
Segregation, Mendel's Law of, 167
Segregation of nuclear determinants., 237
Selection, 3, 78, 89, 92
Selective force, 84
Septum, 22, 23
Sequencing the human genome, 254
Smith, Adam, 82, 83
Sociobiology: the new synthesis, 265
Solute, 128
Solvent, 128
Somatic cells, 198
  differentiation of, 238
"Spaghetti" model, for bacterial recombina-
  tion, 246
Speciation, 93
Species formation, 39
Species problem, 20
Specificity, 155, 156
St. Thomas Aquinas, 31
Stahl, Franklin, 217
Steinbach, Burr, 184
Steno, Nicolaus, 63

Strasburger, Edouard, 173
Streptococcus pneumoniae, 191
Struggle for existence, 83, 87, 89
Sturtevant, Alfred H., 175
Substrate, 155
Succession of worlds, 58
Sutton, Walter S., 173

T

Tatum, E. L., 215
Taxonomy, 17, 43
Technology, 250
Teleology, 69–70
Termination codons, 232
Terminology, genetic, 170
Thales, 36
Thermodynamics, Laws of, 141
Thymine, 187
Tissue doctrine, 109
Toulmin, Stephen, 87
Traits, 165
Transcript, 227
   primary, 225
Transcription, 225
   enzymes involved in, 225
   factors, 226, 238
Transfer RNA, 229
Transformation, 48, 193
   as example of recombination, 244
   in bacteria, 193
Transforming factor, 193, 196
Translation, 232
Transmission genetics, 165, 173
Triplet nature of genetic code, 220
tRNA, 229
Truth, 11
Tschermak, E., 166

U

Ultraviolet light, as mutagen, 236
Uniformitarianism, 56
Uniformity, 56
Unifying principles, 104
Unit characters, 181
Unity, 18
   compared to diversity, 184
Universal Deluge, 55
Unlimited mutability, 54
Uracil, 187

V

Vaccine, 191
Variation, 89
   explanation for, 97
Vectors, in biotechnology, 252
Vendrely, Colette, 198
Vendrely, Roger, 198
Vesalius, Andreas, 21, 23
Vitalism, defined, 106

W

Wallace, Alfred R., 94
Watchmaker God, 71
Water molecule, 122
   polarity of, 126
Watson, James, 200, 202, 255
   in Human Genome Project, 255
Wealth of Nations, 82
Weismann, August
   germ plasm theory, 237
   nuclear determinants, 237
What is life? 105
White blood cells, 186
Wilkins, Maurice, 204, 211
Wilson, E. O., 265
Wilson, Edmund B., 173
Wöhler, Friederich, 106
Wollman, Elie, 246
   bacterial recombination, 246
Women scientists, 260
Woodward, John, 55
World War II, 196

X

X-ray diffraction, 202
X-rays, and mutations, 216